Advanced Numerical Methods with Matlab® 1

Mathematical and Mechanical Engineering Set
coordinated by
Abdelkhalak El Hami

Volume 6

Advanced Numerical Methods with Matlab® 1

Function Approximation and System Resolution

Bouchaib Radi
Abdelkhalak El Hami

iSTE

WILEY

First published 2018 in Great Britain and the United States by ISTE Ltd and John Wiley & Sons, Inc.

ISTE Ltd
27-37 St George's Road
London SW19 4EU
UK

www.iste.co.uk

John Wiley & Sons, Inc.
111 River Street
Hoboken, NJ 07030
USA

www.wiley.com

Library of Congress Control Number: 2018930641

British Library Cataloguing-in-Publication Data
A CIP record for this book is available from the British Library
ISBN 978-1-78630-235-9

Contents

Preface

Most physical problems can be expressed in the form of mathematical equations (e.g. differential equations, integral equations). Historically, mathematicians had to find analytic solutions to the equations encountered in engineering and related fields (e.g. mechanics, physics, biology). These equations are sometimes highly complex, requiring significant work to be simplified. However, in the mid-20th Century, the introduction of the first computers gave rise to new methods for solving equations: numerical methods. This new approach allows us to solve the equations that we encounter (when constructing models) as accurately as possible, thereby enabling us to approximate the solutions of the problems that we are studying. These approximate solutions are typically calculated by computers using suitable algorithms.

Practical experience has shown that, compared to standard numerical approaches, a carefully planned and optimized methodology can improve the speed of computation by a factor of 100 or even higher. This can transform a completely unreasonable calculation into a perfectly routine computation, hence our great interest in numerical methods! Clearly, it is important for researchers and engineers to understand the methods that they are using and, in particular, the limitations and advantages associated with each approach. The computations needed by most scientific fields require techniques to represent functions as well as algorithms to calculate derivatives and integrals, solve differential equations, locate zeros, find the eigenvectors and eigenvalues of a matrix, and much more.

The objective of this book is to present and study the fundamental numerical methods that allow scientific computations to be executed. This involves implementing a suitable methodology for the scientific problem at hand, whether derived from physics (e.g. meteorology, pollution) or engineering (e.g. structural mechanics, fluid mechanics, signal processing).

This book is divided into three parts, with two appendices. Part 1 introduces numerical processing by reviewing a few basic notions of linear algebra. Part 2 discusses how to approximate functions, in three chapters: numerical interpolation, differentiation and integration. Part 3 presents various methods for solving linear systems: direct methods, iterative methods, the method of eigenvalues and eigenvectors and, finally, the method of least-squares.

Each chapter starts with a brief overview of relevant theoretical concepts and definitions, with a range of illustrative numerical examples and graphics. At the end of each chapter, we introduce the reader to the various *Matlab* commands for implementing the methods that have been discussed. As is often the case, practical applications play an essential role in understanding and mastering these methods. There is little hope of being able to assimilate them without the opportunity to apply them to a range of concrete examples. Accordingly, we will present various examples and explore them with *Matlab*. These examples can be used as a starting point for practical exploration.

Matlab is currently widely used in teaching, industry and research. It has become a standard tool in various fields thanks to its integrated toolboxes (e.g. optimization, statistics, control, image processing). Graphical interfaces have been improved considerably in recent versions. One of our appendices is dedicated to introducing readers to *Matlab*.

Bouchaib RADI
Abdelkhalak EL HAMI
January 2018

PART 1

Introduction

1

Review of Linear Algebra

We will denote the fields of real and complex numbers by $\mathbb{R}$ and $\mathbb{C}$ respectively. If there is no need to distinguish between them, we will instead simply refer to the field $\mathbb{K}$ of scalars. The set E, equipped with the two operations of addition and scalar multiplication, denotes a vector space over $\mathbb{K}$ (or a $\mathbb{K}$-vector space).

1.1. Vector spaces

1.1.1. *General definitions*

DEFINITION.– *A vector space over the field* $\mathbb{K}$ *is a set* E *equipped with the two following operations:*

– addition, which equips E *with the structure of a commutative group;*

– an "outer" product of an element of E *by an element of* $\mathbb{K}$*, satisfying the following properties:*

- $\forall \lambda, \mu \in \mathbb{K}, \forall x \in E : (\lambda\mu)x = \lambda(\mu x)$;

- $\forall \lambda, \mu \in \mathbb{K}, \forall x \in E : (\lambda + \mu)x = \lambda x + \mu x$;

- $\forall \lambda \in \mathbb{K}, \lambda(x + y) = \lambda x + \lambda y$;

- $\forall x \in E : 1x = x$ *(where* 1 *is the identity element of* $\mathbb{K}$*).*

The elements of E *are called vectors.*

DEFINITION.– *Let* F *be a subset of the vector space* E*.* F *is a vector subspace of* E *if it is closed under the operations of* E*. In other words:*

– $\forall x, y \in F, x + y \in F$;

– $\forall \lambda \in \mathbb{K}, \forall x \in F, \lambda x \in F$.

DEFINITION.– *Let E be a $\mathbb{K}$-vector space, and suppose that F and G are two vector subspaces of E.*

1) The sum of F and G, written as $F + G$, is defined as the set:

$$F + G = \{x = y + z,\ y \in F,\ z \in G\}.$$

This is a vector subspace of E.

2) $F + G$ is said to be a direct sum if $F \cap G = \{0\}$. If so, we write this sum as $F \oplus G$.

3) If we also have that $E = F \oplus G$, we say that F and G are supplementary subspaces.

THEOREM.– *Suppose that F and G are supplementary. Then, for every element x of E, there exists a unique pair (y, z) in $F \times G$ such that $x = y + z$.*

1.1.2. *Free families, generating families and bases*

DEFINITION.– *Let $B = \{x_1, \dots, x_p\}$ be a family of vectors in E.*

– We say that B is related if one of its vectors is a linear combination of the others, i.e.:

$$\exists (t_1, \dots, t_p) \in \mathbb{K}^p,\ (t_1, \dots, t_p) \neq (0, \dots, 0) \text{ such that } \sum_{i=1}^{p} t_i x_i = 0.$$

– We say that B is free if it is not related, in which case its vectors are said to be linearly independent.

– We say that B is a generating family of E (or generates E) if every element of E is a linear combination of the elements of B.

DEFINITION.– *A family $B = \{e_1, \dots, e_p\}$ of elements in a vector space E is said to be a basis of E if it is free and generates E.*

The canonical basis is one particular example of a basis, which is defined as follows:

DEFINITION.– *The canonical basis is the basis of vectors $\{e_i\}_{i=1\dots n}$ such that the j-th element of e_i is 0 except when $i = j$, in which case it is equal to 1.*

Thus, every vector x in $\mathbb{R}^n$ may be decomposed with respect to the canonical basis as follows:

$$\forall x \in \mathbb{R}^n, \; x = \sum_{i=1}^{n} x_i e_i.$$

THEOREM.– *In a vector space generated by a finite family of elements, every basis has the same number of elements.*

DEFINITION.– *The dimension of a vector space E generated by a finite family is defined as the number of elements in any given basis of E. This value is denoted as "dim E".*

In any vector space with finite dimension n, we always use the same basis, $B = \{e_1, \ldots, e_n\}$. Thus, each vector x of E may be uniquely decomposed with respect to B as follows:

$$x = x_1 e_1 + \cdots + x_n e_n.$$

The element $X = (x_1, \ldots, x_n)$ in $\mathbb{K}^n$ may therefore be unambiguously chosen as a representation of x.

THEOREM.– *Let E be an n-dimensional vector space, and suppose that F and G are two vector subspaces of E. Then:*

1) every free family of n vectors is a basis;

2) every generating family of n vectors is a basis;

3) dim $F \leq$ dim E;

4) if $F \cap G = \{0\}$, then dim $F+$ dim $G \leq$ dim E;

5) in particular, if $E = F \oplus G$, then dim $F+$ dim $G =$ dim E.

1.2. Linear mappings

DEFINITION.– *Let E and F be two vector spaces over the field K. A mapping $u\colon E \to F$ is said to be a linear mapping if it satisfies the following properties:*

– $u(x + y) = u(x) + u(y)$ $\forall x, y \in E$;

– $u(\lambda x) = \lambda u(x)$ $\forall x \in E$, $\forall \lambda \in K$.

The set of linear mappings from E to F is denoted as $\mathcal{L}(E, F)$.

The linear mappings are the mappings that preserve the vector space structure.

DEFINITION.–

1) The kernel of u*, written as Ker(*u*), is the vector subspace of* E *defined by:*

$$\mathrm{Ker}(u) = \{x \in E \mid u(x) = 0\}.$$

2) The image of u*, written as Im(*u*), is the vector subspace of* F *defined by:*

$$\mathrm{Im}(u) = \{y \in F \mid \exists x \in E \text{ such that } y = u(x)\}.$$

THEOREM.–

– u *is injective if and only if Ker(*u*)* $= \{0\}$.

– u *is surjective if and only if Im(*u*)* $= F$.

DEFINITION.– *Let* I_E *(respectively* I_F*) be the identity mapping of* E *(respectively* F*). The linear mapping* u *from* E *to* F *is said to be invertible if there exists a linear mapping* u^{-1} *from* F *to* E *such that:*

$$u^{-1} \circ u = i_E \quad \text{and} \quad u \circ u^{-1} = i_F. \qquad [1.1]$$

It follows that every invertible linear mapping is bijective, i.e. injective and surjective.

THEOREM.– *Let* $u \in \mathcal{L}(E, E)$*. The following are equivalent:*

– u *is injective;*

– u *is surjective;*

– u *is bijective.*

THEOREM.– *Let* $u \in \mathcal{L}(E, F)$ *and suppose that* $B = \{e_1, \ldots, e_n\}$ *is a basis of* E*. Then:*

– *if* u *is injective,* $\{u(e_1), \ldots, u(e_n)\}$ *is a basis of Im(*u*);*

– *if* u *is surjective,* $\{u(e_1), \ldots, u(e_n)\}$ *is a generating family of* F*;*

– *the following relation holds:*

$$\dim E = \dim \mathrm{Ker}(u) + \dim \mathrm{Im}(u).$$

DEFINITION.– *The rank of a linear mapping, denoted as "rank u", is the dimension of Im(u).*

1.3. Matrices

In this section, E, F and G are three vector spaces over the field $\mathbb{K}$, with finite dimensions n, p and q respectively. The families $B_E = \{e_1, \dots, e_n\}$, $B_F = \{f_1, \dots, f_p\}$ and $B_G = \{g_1, \dots, g_q\}$ are the bases of E, F and G.

DEFINITION.– *Let $u \in \mathcal{L}(E, F)$. The matrix of u with respect to the bases B_E and B_F is defined as an array A of scalars (i.e. elements of $\mathbb{K}$) with p rows and n columns such that the j-th column of A is given by the components of the vector $u(e_j)$ with respect to the basis B_F.*

If a_{ij} is the element of A at the intersection of the i-th row and the j-th column, then:

$$u(e_j) = \sum_{i=1}^{p} a_{ij} f_i. \qquad [1.2]$$

The matrix A, which has p rows and n columns, is said to be of format or type (p, n), or is called a $p \times n$ (p-by-n) matrix.

It does not make much sense to prove results on matrices without referring to the linear mappings that they represent. We will use this link between mappings and matrices to define operations on matrices.

1.3.1. *Operations on matrices*

DEFINITION.– *Let A and B be two $p \times n$ matrices. The sum $A + B$ of A and B is the $p \times n$ matrix C with coefficients c_{ij} defined by:*

$$c_{ij} = a_{ij} + b_{ij}.$$

The matrix C thus obtained is the matrix of the linear mapping obtained by summing the two linear mappings represented by A and B.

Similarly, we define the product of a scalar λ and a matrix A as the matrix λA obtained by multiplying each coefficient of A by λ. The set of matrices with p rows and n columns is a vector space, written as $\mathcal{M}_{p,n}(\mathbb{K})$ (or simply $\mathcal{M}_{p,n}$ when the underlying field is explicit).

DEFINITION.– *Let* $u \in \mathcal{L}(F, G)$ *and* $v \in \mathcal{L}(E, F)$. *Define* $w = u \circ v$ *(and therefore* $v \in \mathcal{L}(E, G)$*). Let* A *be the matrix of* u *with respect to the bases* B_F *and* B_G*, and* B *the matrix of* v *with respect to the bases* B_E *and* B_F*. By definition, the matrix of* w *with respect to the bases* B_E *and* B_G *is equal to the product of* A *and* B*, written as* AB.

THEOREM.– *Let* $A \in \mathcal{M}_{n,p}$ *and* $B \in \mathcal{M}_{p,q}$*, and define* $C = AB$*. Then,* $C \in \mathcal{M}_{n,q}$ *and its elements are given by the formula:*

$$c_{ij} = \sum_{k=1}^{p} a_{ik} b_{kj} \quad i = 1, \ldots, n, \; j = 1, \ldots, q. \qquad [1.3]$$

DEFINITION.– *The matrix* A *is said to be invertible (or regular or non-singular) if the linear mapping associated with* A *is invertible. If* $A \in \mathcal{M}_{n,n}$ *is invertible, then there exists an inverse matrix, denoted as* A^{-1}*, such that:*

$$AA^{-1} = A^{-1}A = I,$$

where I *is the identity matrix in* $\mathcal{M}_{n,n}$.

Properties

If the matrices A, B and C have dimensions that allow them to be multiplied, then, given any $\alpha \in \mathbb{R}$, the following relations hold:

- $C(A + B) = CA + CB$;
- $(A + B)C = AC + BC$;
- $A(BC) = (AB)C = ABC$;
- $\alpha(AB) = (\alpha A)B = A(\alpha B)$.

1.3.2. *Change-of-basis matrices*

DEFINITION.– *Let* E *be an* n*-dimensional vector space equipped with a basis* $B_E = \{e_1, \ldots, e_n\}$*. Consider another basis of E, denoted as* $B'_E = \{e'_1, \ldots, e'_n\}$*. The change-of-basis matrix is defined as the matrix* P *whose columns are the components of the elements of* B'_E *with respect to the original basis* B_E*. Therefore:*

$$e'_j = \sum_{i=1}^{n} p_{ij} e_i.$$

We cannot directly calculate the new components of a vector $\vec{x}$ as a function of its original components using the change-of-basis matrix. Suppose that:

$$x = \sum_{i=1}^{n} x_i e_i = \sum_{i=1}^{n} x'_i e'_i.$$

Let X be the vector in $\mathbb{K}^n$ formed by the components of x with respect to the original basis B_E, and write X' for the vector in $\mathbb{K}^n$ formed by the components of x with respect to the new basis B'_E. It can be shown that:

$$X = PX',$$

or, in other words, noting that change-of-basis matrices are always invertible:

$$X' = P^{-1}X.$$

THEOREM.– *Suppose that $u \in \mathcal{L}(E, E)$ has the matrix A with respect to the basis B_E. Then, the matrix A' of u with respect to the basis B'_E satisfies the following formula:*

$$A' = P^{-1}AP.$$

Change-of-basis operations are extremely important in practice, since they allow square matrices to be expressed in more "favorable" forms (diagonal, upper triangular or lower triangular structure).

1.3.3. *Matrix notations*

For any $A \in \mathcal{M}_{p,n}$, we will adopt the following notation, unless otherwise stated:

– A_j is the j-th column of the matrix A, which can therefore be represented by:

$$A = (A_1, A_2, \ldots, A_n).$$

– A^T is the transpose of the matrix A (obtained by switching the rows and columns of A). Thus, $A^T \in \mathcal{M}_{n,p}$ and:

$$(A^T)_{ij} = a_{ji} \quad \forall i, j, \; 1 \leq i \leq p, 1 \leq j \leq n.$$

– If $A \in \mathcal{M}_{np}(\mathbb{C})$, we write $\overline{A}$ for the complex conjugate matrix of A (whose coefficients are the complex conjugates of the coefficients of A).

– If A is a square matrix ($p = n$), we define the trace of A to be the sum of its diagonal elements:

$$\text{tr}(A) = \sum_{i=1}^{n} a_{ii}.$$

– We write $D = \text{diag}(d_1, \dots, d_n)$ for the diagonal matrix D.

– Every vector in $\mathbb{K}^n$ may be identified with an $n \times 1$ matrix (with n rows and one single column). If X is a column vector, then X^T is a row vector (a $1 \times n$ matrix, with one single row and n columns).

DEFINITION.– *Any matrix $A \in \mathcal{M}_{n,n}$ satisfying:*

$$a_{ij} = a_{ji} \quad \forall i, j \in \{1, \dots, n\}$$

is called a symmetric matrix, in which case $A^T = A$.

1.4. Determinants

To define the notion of determinant, we require the notions of permutation and signature, so we will present them first.

DEFINITION.– *Let $I_n = \{1, 2, \dots, n\}$. A permutation is a bijective mapping from I_n to itself. A transposition is a permutation that only exchanges two consecutive elements (we write τ_j for the permutation that exchanges j and $j+1$).*

We write $\mathcal{P}_n$ for the set of permutations of I_n. Every permutation may be written as a product of transpositions. This product is not unique, but the number of transpositions in any given decomposition always has the same parity.

DEFINITION.– *The signature of a permutation σ is defined as the value sign(σ), which is $+1$ if the permutation may be decomposed into an even number of transpositions, and -1 otherwise.*

DEFINITION.– *The determinant of a matrix $A \in \mathcal{M}_{nn}$, written as det(A), is the number:*

$$\det(A) = \sum_{\sigma \in \mathcal{P}} \text{sign}\,(\sigma) a_{\sigma(1),1} a_{\sigma(2),2} \cdots a_{\sigma(n),n}. \qquad [1.4]$$

There are only finitely many permutations of a finite set, and the cardinal of $\mathcal{P}_n$ is finite. Therefore, the sum in formula [1.4] is finite.

Properties

The most important properties of determinants are as follows:

1) $\det(\overline{A}) = \overline{\det(A)}$;

2) $\det(A) = \det(A^T)$;

3) $\forall \sigma \in \mathcal{P}_n$, $\det(A_{\sigma(1)}, A_{\sigma(2)}, \ldots, A_{\sigma(n)}) = \text{sign}(\sigma)\det(A)$.

4) The mapping that sends any given column of A to the determinant of A is linear. In particular, if one of the columns of A is a linear combination of the others, then $\det(A) = 0$.

The mapping that sends the columns of the matrix A to the determinant of A is an alternating multilinear mapping. The multilinearity of this mapping follows from the aforementioned property 4. Hence:

$$\det(\alpha A) = \alpha^n \det(A).$$

This mapping is said to be an alternating mapping because, by the aforementioned property 3, permuting any two consecutive columns of the matrix changes the sign of the determinant.

Properties

1) In general, $\det(A + B) \neq \det(A) + \det(B)$.

2) Let A and B be two matrices in $\mathcal{M}_{nn}$. Then:

$$\det(AB) = \det(A)\det(B).$$

Let $A \in \mathcal{M}_{nn}(K)$. We write $A_{|i,j|} \in \mathcal{M}_{n-1,n-1}(K)$ for the minor of A obtained by deleting the i-th row and the j-th column. In "practice", determinants are usually calculated using the following result:

THEOREM.– *The determinant can be calculated by expanding along a column:*

$$\forall j \in \{1, \ldots, n\} \quad det(A) = \sum_{i=1}^{n} a_{ij}(-1)^{i+j} det(A_{|i,j|}). \qquad [1.5]$$

Alternatively, we can also expand along rows using an analogous formula, by the aforementioned property 2.

DEFINITION.– *The scalar* $(-1)^{i+j}$ det $(A_{|i,j|})$ *is said to be the cofactor of the element* a_{ij}. *The matrix of cofactors is called the comatrix, which is written as* $co(A)$.

THEOREM.– *A square matrix A is invertible (or regular or non-singular) if and only if its determinant is non-zero. If so, the inversion formula may be written as:*

$$A^{-1} = \frac{1}{det(A)}[co(A)]^T. \tag{1.6}$$

1.5. Scalar product

DEFINITION.– *A scalar product on E is a mapping from $E \times E$ to $\mathbb{R}$:*

$$(\vec{x}, \vec{y}) \rightarrow <\vec{x}, \vec{y}> = (\vec{x}, \vec{y})_E$$

satisfying the following properties:

– $<\vec{x}, \vec{x}> \geq 0$;

– $<\vec{x}, \vec{x}> = 0 \Rightarrow \vec{x} = 0$;

– $<\vec{x}, \vec{y}> = <\vec{y}, \vec{x}>$;

– $<\vec{x}, \vec{y} + \vec{z}> = <\vec{x}, \vec{y}> + <\vec{x}, \vec{z}>$;

– $<\vec{x}, \alpha\vec{y}> = \alpha <\vec{x}, \vec{y}> \forall \alpha \in \mathbb{R}$.

LEMMA.– *The scalar product is a symmetric bilinear form on E.*

1.6. Vector norm

The norm of a vector may be defined in several different ways. The most common is, of course, the Euclidean norm of a vector $\vec{x} = (x_1, x_2, \ldots, x_n)^T$:

$$\|\vec{x}\| = \sqrt{<\vec{x}, \vec{x}>} = \sqrt{\sum_{i=1}^{n} x_i^2}, \quad x_i \in \mathbb{R}.$$

Geometrically, this norm represents the length of the vector. However, norms are defined more generally.

DEFINITION.– *The mapping $\|.\|: E \rightarrow \mathbb{R}$ (here, E is a real vector space) is said to be a vector norm if the following properties hold:*

1) $\| \vec{x} \| \geq 0, \forall \vec{x} \in E$ *and* $\| \vec{x} \| = 0 \Leftrightarrow \vec{x} = 0$;

2) $\| \lambda\vec{x} \| = |\lambda| \| \vec{x} \|, \forall a \in \mathbb{R}, \vec{x} \in E$;

3) $\| \vec{x} + \vec{y} \| \leq \| \vec{x} \| + \| \vec{y} \|, \forall \vec{x}, \vec{y} \in E$.

It can easily be verified that the Euclidean norm defined above satisfies these properties. As mentioned above, we can define several different norms on the same vector space. One well-known example of norms on $\mathbb{R}^n$ is the family of p-norms. Given any vector $\vec{x}$, the p-norms are defined as follows:

$$\| \vec{x} \|_p = \left(\sum_{i=1}^{n} |x_i^p| \right)^{1/p}.$$

When $p = 2$, we recover the definition of the Euclidean norm. Another example is the infinity norm, which is defined as follows:

$$\|\vec{x}\|_\infty = \max_{1 \leq i \leq n} |x_i|.$$

The infinity norm is one of the most widely used norms, together with the 2-norm and the 1-norm. Other more exotic norms can also be defined (e.g. elliptic norms); however, we will simply work with the norms defined above.

EXAMPLE.– The 1-norm, 2-norm and ∞-norm of the vector $\vec{x} = (1, 0, 1, 4)^T$ may be calculated as follows:

$$\|\vec{x}\|_1 = \sum_{i=1}^{4} |x_i| = 6;$$

$$\|\vec{x}\|_2 = \sqrt{\sum_{i=1}^{4} x_i^2} = \sqrt{18} = 3\sqrt{2};$$

$$\|\vec{x}\|_\infty = \max_{1 \leq i \leq 4} | x_i | = 4.$$

This example shows that different norms can take different values.

1.7. Matrix eigenvectors and eigenvalues

1.7.1. *Definitions and properties*

Throughout the rest of this section, E denotes a vector space over $\mathbb{C}$.

DEFINITION.– *Let $u \in \mathcal{L}(E, E)$. We say that $\lambda \in \mathbb{C}$ is an eigenvalue of u if there exists a non-zero vector $\vec{y}$ such that:*

$$u(\vec{y}) = \lambda \vec{y}. \qquad [1.7]$$

The vector $\vec{y}$ is said to be an eigenvector of the eigenvalue λ.

With respect to any given basis of E, the linear mapping u may be represented by a matrix $A \in \mathcal{M}_{nn}$, and the eigenvector y may be represented by an element Y of $\mathbb{C}^n$. Therefore:

$$AY = \lambda Y.$$

This allows us to introduce the following definition:

DEFINITION.– *Let $A \in \mathcal{M}_{nn}(\mathbb{C})$. We say that $\lambda \in \mathbb{C}$ is an eigenvalue of A if there exists a non-zero $Y \in \mathbb{C}^n$ such that:*

$$AY = \lambda Y. \tag{1.8}$$

The vector Y is said to be an eigenvector of A for the eigenvalue λ.

THEOREM.– *The complex number λ is an eigenvalue of A if and only if:*

$$det(\lambda I - A) = 0. \tag{1.9}$$

If A is not invertible, then 0 is an eigenvalue of A.

DEFINITION.– *The characteristic polynomial of A, written as $P_A(s)$, is the polynomial:*

$$P_A(s) = det(sI - A) = s^n + \mu_1 s^{n-1} + \cdots + \mu_{n-1} s + \mu_n. \tag{1.10}$$

THEOREM.– *By d'Alembert's theorem on the roots of a polynomial, every $n \times n$ square matrix has n eigenvalues (counted with multiplicity). By writing $\lambda_1, \ldots, \lambda_p$ for the p distinct roots of the polynomial and n_i for the multiplicity of λ_i, it follows that:*

$$P_A(s) = (s - \lambda_1)^{n_1}(s - \lambda_2)^{n_2} \ldots (s - \lambda_p)^{n_p}. \tag{1.11}$$

Note that the characteristic polynomial is fully determined by the eigenvalues of A, which are also the eigenvalues of the linear mapping represented by this matrix. Therefore:

DEFINITION.– *Let $u \in \mathcal{L}(E, E)$. The characteristic polynomial of u, written as P_u, is defined as the characteristic polynomial of any matrix representing u with respect to any given basis of E.*

THEOREM.– *Let $u \in \mathcal{L}(E, E)$. The trace of u, written as $tr(u)$, is defined by either of the following two equivalent conditions:*

– *$tr(u) = tr(A)$, where A is an arbitrary matrix representing u;*

– $tr(u) = \sum_{i=1}^{n} \lambda_i$, *where* λ_i, $i = 1, \ldots, n$ *are the* n *(possibly non-distinct) eigenvalues of* u.

THEOREM.– *Let* y *be an eigenvector of* A *with the eigenvalue* λ. *Then, equivalently:*

– λ *is an eigenvalue of* A^T *(however, the corresponding eigenvector is not* y *in general);*

– $\bar{\lambda}$ *is an eigenvalue of* $\bar{A}$ *and* $\bar{y}$ *is the corresponding eigenvector;*

– *if* $R \in \mathcal{M}_{n,n}(K)$ *is invertible, then* λ *is an eigenvalue of* $R^{-1}AR$.

1.7.2. *Matrix diagonalization*

Let $u \in \mathcal{L}(E, E)$.

THEOREM.– *Let* A *be a matrix, and suppose that* $\mathcal{Y} = \{Y_1, \ldots, Y_n\}$ *is a basis of eigenvectors of* A *with possibly non-distinct eigenvalues* $\lambda_1, \ldots, \lambda_n$. *Let* T *be the change-of-basis matrix from the canonical basis of* $\mathbb{C}^n$ *to* $\mathcal{Y}$. *Consider the matrix* D *defined by:*

$$D = T^{-1}AT. \qquad [1.12]$$

Then, D *is a diagonal matrix with eigenvalues along the diagonal. We write that:*

$$D = diag(\lambda_1, \ldots, \lambda_n). \qquad [1.13]$$

THEOREM.– *If the eigenvalues of* A *are all distinct, then their eigenvectors form a basis.*

DEFINITION.– *Two square matrices* A *and* B *are said to be similar if there exists an invertible matrix* U *such that:*

$$B = U^{-1}AU.$$

A matrix is said to be diagonalizable if it is similar to a diagonal matrix, or triangularizable if it is similar to a triangular matrix.

1.7.3. *Triangularization of matrices*

In the case where there does not exist a basis of eigenvectors, we can still write the matrix of u in a triangular form in $\mathbb{C}$ by performing a change of basis.

THEOREM.– *For every $u \in \mathcal{L}(E, E)$, there exists a basis of E with respect to which the matrix T of u is triangular, with the eigenvalues of u along the diagonal.*

The above theorem states that we can use a change of basis to express every matrix in a triangular form in $\mathbb{C}$. This does not hold in general in $\mathbb{R}$.

1.8. Using *Matlab*

Matlab allows us to define arrays or matrices of complex numbers in the same way as arrays or matrices of real numbers:

```
>>Z=[1+2i 0 z1;1 z2 z1+z2;z1*z2 z1-z2 z1/z2]
Z =
    1.0000 + 2.0000i    0                   3.5000 - 1.2500i
    1.0000              1.3140 - 0.0948i    4.8140 - 1.3448i
    4.4805 - 1.9744i    2.1860 - 1.1552i    2.7181 - 0.7551i
```

We can also perform arithmetic operations on matrices:

```
>>A=Z^(-1)
A =
     0.2279 - 0.1872i   -0.2475 + 0.1665i    0.1564 - 0.0282i
    -0.6237 + 0.3429i    0.4146 - 0.4741i    0.0537 + 0.3314i
     0.1251 - 0.0321i    0.1174 + 0.1358i   -0.0282 - 0.0914i
>>A*Z
ans =
     1.0000 + 0.0000i    0.0000 + 0.0000i    0.0000 - 0.0000i
    -0.0000 - 0.0000i    1.0000              0.0000 + 0.0000i
     0.0000 - 0.0000i   -0.0000              1.0000
>>A/Z
ans =
     0.1328 - 0.2826i   -0.0267 + 0.2886i   -0.0451 - 0.1223i
    -0.1566 + 0.6724i    0.0056 - 0.5356i    0.1202 + 0.1689i
    -0.1037 - 0.0857i    0.0965 + 0.0148i   -0.0276 + 0.0428i
```

A matrix of order 0 is a scalar, and a matrix of order 1 is a vector. This vector can be either a row vector or a column vector.

```
>>x=[1,2,3,4]
x =
    1 2 3 4
```

Alternatively:

```
>>x=[0 1 2 3]
```

also returns a row vector. For example, $x(2) = 1$.

```
>>x=[1 ;2 ;3 ;4]
x =
     1
     2
     3
     4
```

```
>>x=[0 ;1 ;2 ;3]'
```

returns a row vector.

```
>>x=[0 1 2 3]'
```

returns a column vector.

REMARK.– In *Matlab*, the row and column indices always start at 1 rather than 0, unlike in other programming languages. Thus, $x(0)$ is not defined.

A matrix of order greater than 1 is a two-dimensional matrix. For example:

```
>>x=[0 :2 ;4 :6]
x =
     0 1 2
     4 5 6
```

This is a matrix with 2 rows and 3 columns.

```
>>y=[0 :2 ;4 :6]'
y =
     0   4
     1   5
     2   6
```

The size of the matrix y is returned by the *size* function:

```
>>size(y)
ans =
     3   2
```

The answer returned is 3 rows and 2 columns. The j-th column of the matrix x is returned by *y(:,j)*. For example, when $j = 2$, we have *y(:,2)= 4 5 6*. The i-th row of the matrix x is returned by *y(i,:)*. For $i = 2$, we have *y(2,:)= 1 5*.

Given an $n \times n$ square matrix A, the corresponding identity matrix is returned by the *eye* function. For example, for $n = 3$, we have:

```
>>A=[1:3;4:6;7:9]
A =
     1   2   3
     4   5   6
     7   8   9
>>eye(size(A))
ans =
     1   0   0
     0   1   0
     0   0   1
```

eye(size(A)) returns the identity matrix with the same dimensions as A.

Consider the row matrix (row vector) *x=[2 15 0]*. The function *sort(x)* returns a row matrix whose elements have been sorted into increasing order:

```
>>sort(x)
ans =
     0   2  15
```

sort(x') returns a column matrix whose elements have been sorted into increasing order:

```
>>sort(x')
ans =
     0
     2
     15
```

sum(x) calculates the sum of the column elements of the matrix x:

```
>>sum(x)
ans =
     17
>>sum([4 2 1 0;8 9 3 12])
ans =
     12 11 4 12
```

To find the maximum and minimum elements of the vector x, we can use the functions *max* and *min*:

```
>>max(x)
ans =
    15
>>min(x)
ans =
    0
```

The *max* and *min* commands can also be applied to any arbitrary matrix y.

2

Numerical Precision

2.1. Introduction

Implementing a numerical method on a machine introduces a few additional difficulties, since we need to use approximations to represent numbers.

Computer-based representations of numbers rely on two types of convention:

– a writing convention: we must decide how many characters should be used to represent the number (which is known as the external representation);

– an encoding convention: each character of the external representation is encoded into bits (0 or 1).

The study of numerical errors is an important aspect of scientific computation. Errors can be introduced into the solutions of a problem by multiple sources. Data rounding operations are the primary source of error in computer-based calculations, followed by operations on floating-point real numbers. Rounding errors are caused by the fact that the computer can only recognize finitely many numbers, such as those with finitely many non-zero digits after the decimal point, but numbers such as $\frac{1}{3}$ or $\sqrt{2}$ have infinite non-zero digits after the decimal point.

Once an error has been introduced, it usually propagates throughout the whole calculation. This motivates the concept of numerical stability.

2.2. Machine representations of numbers

Computers use a floating-point system to represent numbers.

DEFINITION.– *Let x be a non-zero real number. The floating-point representation of x is:*

$$x = \pm 0.a_1 \ldots a_N \times b^E,$$

where $b \in \mathbb{N}^*$ is said to be the base, $a = 0.a_1 \ldots a_N$ is said to be the mantissa, $0 \leq a_i < b$, $a_1 \neq 0$, $E \in \mathbb{Z}$ is the exponent, which takes values between m and M ($-m \leq E \leq M$), and $N \in \mathbb{N}$ is the number of significant figures.

If $E < -m$ or $E > M$, the real number x does not have a floating-point representation in this system, in which case the computer does not know the number. This is called "underflow" (if $E < -m$) or "overflow" (if $E > M$) [KIU 05].

EXAMPLE.– Consider the numbers:

$$a = 0.005673\,;\; b = -987.5678\,;\; c = 87654900\,;\; d = 345.876542.$$

Their floating-point representations are:

$$a = 0.5673 \times 10^2\,;\; b = -0.9875678 \times 10^3\,;\; c = 0.876549 \times 10^8\,;$$

$$d = 0.345876542 \times 10^3.$$

DEFINITION.– *Let x be a real number with floating-point representation $x = \pm 0.a_1 \ldots a_N a_{N+1} \times b^E$. If the machine is only working with N significant figures, we need to define the rounded value $Round(x)$ of x:*

– *if $a_{N+1} \geq \frac{b}{2}$, then $Round(x) = \pm 0.a_1 \ldots a_N \times b^E + 0.\underbrace{0\ldots0}_{N-1}1 \times b^E$;*

– *if $a_{N+1} < \frac{b}{2}$, then $Round(x) = \pm 0.a_1 \ldots a_N \times b^E$.*

The relative error is:

$$\frac{|x - Round(x)|}{|x|} \leq \underbrace{\frac{b}{2} b^{-N}}_{\text{machine precision}}.$$

EXAMPLE.– We will perform the following four operations by both truncating and rounding to four significant figures:

1) $x_1 \oplus x_2$ with $x_1 = 56170$ and $x_2 = 285.1$;

2) $x_1 \ominus x_2$ with $x_1 = 63170$ and $x_2 = 65.33$;

3) $x_1 \otimes x_2$ with $x_1 = 561.5$ and $x_2 = 38.59$;

4) $x_1 \oslash x_2$ with $x_1 = 26170$ and $x_2 = 385700$.

We find:

1)
$$\begin{aligned} x_1 + x_2 &= 0.5617 \times 10^5 + 0.2851 \times 10^3 \\ &= (0.5617 + 0.002851) \times 10^5 \\ &= 0.564551 \times 10^5 \end{aligned}$$

$x_1 \oplus x_2 = 0.5645 \times 10^5$ by truncating.
$x_1 \oplus x_2 = 0.5646 \times 10^5$ by rounding.

2)
$$\begin{aligned} x_1 - x_2 &= 0.6317 \times 10^5 - 0.6533 \times 10^2 \\ &= (0.6317 - 0.0006533) \times 10^5 \\ &= 0.6310467 \times 10^5 \end{aligned}$$

$x_1 \ominus x_2 = 0.6310 \times 10^5$ by truncating and rounding.

3)
$$\begin{aligned} x_1 \times x_2 &= 0.5615 \times 10^3 \times 0.3859 \times 10^2 \\ &= 0.5615 \times 0.3859 \times 10^5 \\ &= 0.21668285 \times 10^5 \end{aligned}$$

$x_1 \otimes x_2 = 0.2166 \times 10^5$ by truncating.
$x_1 \otimes x_2 = 0.2167 \times 10^5$ by rounding.

4)
$$\begin{aligned} x_1 \div x_2 &= 0.2617 \times 10^5 \div 0.3857 \times 10^6 \\ &= (0.2617 \div 0.3857) \times 10^{-1} \\ &= 0.678506611 \times 10^{-1} \end{aligned}$$

$x_1 \oslash x_2 = 0.6785 \times 10^{-1}$ by truncating and rounding.

2.3. Integers

2.3.1. *External representation*

Given an integer b that is strictly greater than 1 (the base), every positive integer may be uniquely written as:

$$N = \sum_{i=0}^{n} a_i b^i \text{ with } 0 \leq a_i < b; \;\; a_n \neq 0.$$

The integers a_i are the "digits" of N in the base b.

2.3.1.1. *Writing conventions*

To write the digits of a number base b, we use the following symbols:

$$0, 1, 2, 3, 4, 5, 6, 7, 8, 9, A, B, C, \ldots$$

2.3.1.2. *Positional conventions*

The number N is written in the base b as $a_n a_{n-1} \ldots a_0$ (with units on the right). Zero is represented by 0.

– If $b = 2$, this representation is said to be binary.

– If $b = 8$, this representation is said to be octal.

– If $b = 16$, this representation is said to be hexadecimal.

EXAMPLE.– The decimal number 29 may be written as:

$$\begin{aligned} N &= 2^4 + 2^3 + 2^2 + 2^0 \rightarrow 11101 && \text{in binary} \\ &= 3 \times 8^1 + 5 \times 8^0 \rightarrow 35 && \text{in octal} \\ &= 16 + 13 \times 16^0 \rightarrow 1D && \text{in hexadecimal.} \end{aligned}$$

2.3.2. *Internal representation of positive integers*

Each digit $a_n, a_{n-1}, \ldots, a_0$ of a number written in the base b is associated with a binary encoding. We can therefore have:

– hexadecimal expressions encoded in binary;

– octal expressions encoded in binary.

EXAMPLE.–

$$29 \rightarrow 1D \rightarrow \underbrace{0001}\underbrace{1101}$$

$$29 \rightarrow 35 \rightarrow \underbrace{011}\,\underbrace{101}$$

Note that if we delete unnecessary zeros in these encodings, hexadecimal encoded in binary = octal encoded in binary = binary.

2.4. Real numbers

2.4.1. *External representation*

Given an integer $b > 1$ (the base), every non-zero real number x may be uniquely written as:

$$x = \pm M b^E,$$

where M (the normalized mantissa) is a real number such that $\frac{1}{b} \leq M < 1$ and E (the exponent) is a positive or negative integer or zero.

If we represent the mantissa in the base b:

$$M = \sum_{i=1}^{\infty} a_i b^{-i} \; 0 \leq a_i < b \; a_1 \neq 0,$$

then the a_i are the digits of M in the base b.

This yields the normalized floating-point expression of x:

$$x_b = \pm 0.a_1 a_2 \ldots a_n \ldots \times b^E.$$

EXAMPLE.– The expression $x_{10} = 425.33$ denotes the real number $x = 4 \times 10^2 + 2 \times 10 + 5 + 3 \times 10^{-1} + 3 \times 10^{-2}$, whereas the expression $x_6 = 425.33$ denotes the real number $x = 4 \times 6^2 + 2 \times 6 + 5 + 3 \times 6^{-1} + 3 \times 6^{-2}$. The same rational number can have infinitely many digits in one base and finitely many digits in another. For example, the fraction $\frac{1}{3}$ has infinitely many digits in base 10, $x_{10} = 0.33333$, but only one digit in base 3, $x_3 = 0.1$.

2.4.2. *Internal encoding of real numbers*

Consider now an integer m that is strictly greater than 1. We can encode any given real number by a triple (S,M,E):

– S: sign $\pm$;

– M: mantissa normalized to m digits;

– E: exponent.

The real number $x = \pm 0.a_1 a_2 \ldots a_n \ldots \times b^E$ has the following floating-point representation:

$$\bar{x} = \pm 0.x_1 x_2 \ldots x_m \times b^E.$$

Each computer has its own specific way of encoding the sign, the mantissa and the exponent.

2.5. Representation errors

In order to represent real numbers with limited space, we must replace each real number x by $\bar{x}$.

The difference $|x - \bar{x}|$ is the absolute representation error. The relative representation error is given by $\left|\dfrac{x-\bar{x}}{x}\right|$. The mantissa with m digits may be deduced from the exact mantissa in two ways:

– truncation: $x_i = a_i$, $i = 1, \ldots, m$;

– rounding: $x_i = a_i$, $i = 1, \ldots, m-1$, and

$$x_m = \begin{cases} a_m & \text{if } a_{m+1} \in [0, \frac{b}{2}[\\ a_m + 1 & \text{if } a_{m+1} \in [\frac{b}{2}, b[\end{cases}$$

Property

$$\left|\frac{x-\bar{x}}{x}\right| \leq \begin{cases} b^{1-m} & \text{for representations obtained by truncation} \\ \dfrac{1}{2}b^{1-m} & \text{for representations obtained by rounding.} \end{cases}$$

For truncation, this can be done by writing M for the true mantissa and $\bar{M}$ for the mantissa with m digits, and then noting that:

$$\left|\frac{x-\bar{x}}{x}\right| = \left|\frac{M-\bar{M}}{M}\right| \leq b|M-\bar{M}| = b\sum_{i=m+1}^{+\infty} a_i b^{-i}$$

$$\leq b(b-1)\sum_{i=m+1}^{+\infty} b^{-i} = b(b-1)\frac{b^{-(m+1)}}{1-b^{-1}} = b^{1-m}.$$

A similar argument can be used to show the claim for the rounding error.

We will now give two examples of errors in elementary operations.

EXAMPLE.– Suppose that we wish to calculate $x - y$ with $x = 1.02$, $y = 0.0617$, $b = 10$ and $m = 3$.

With 0 guard digits:

– denormalization: $x = 0.102 \times 10^1$ and $y = 0.006 \times 10^1$;

– intermediate result: 0.096×10^1;

– normalization: $x - y = 0.96 \times 10^0$.

With 1 guard digit:

– denormalization: $x = 0.1020 \times 10^1$ and $y = 0.0061 \times 10^1$;

– intermediate result: 0.0959×10^1;

– normalization: $x - y = 0.959 \times 10^0$.

EXAMPLE.– Suppose that we wish to calculate $x \times y$ with $x = 0.333$, $y = 0.123$, $b = 10$ and $m = 3$.

With 0 guard digits:

– denormalization: $x = 0.333$ and $y = 0.123$;

– intermediate result: 0.039;

– normalization: $x \times y = 0.39 \times 10^{-1}$.

With 1 guard digit:

– denormalization: $x = 0.3330$ and $y = 0.1230$;

– intermediate result: 0.0408;

– normalization: $x \times y = 0.408 \times 10^{-1}$.

EXAMPLE.– Let $x = 0.124322 \times 10^4$ and $y = 0.123171 \times 10^4$. Computing $x - y$ on a machine with four significant figures yields $Round(x) - Round(y) = 0.1243 \times 10^4 - 0.1231 \times 10^4 = 0.11 \times 10^2$, with only two significant figures remaining! This phenomenon is known as a loss of significance.

2.5.1. *Properties of computer-based arithmetic*

Machine operations do not necessarily satisfy the properties of associativity, distributivity and commutativity.

EXAMPLE.– Let $x = 10^{20}$, $y = -10^{20}$, $z = 1$. It can be verified that $(x + y) + z = 1$, but $x + (y + z) = 0$.

We need to distinguish between the mathematical operation $\bullet$ and the computer-based operation $\odot$. Therefore, we will write:

$c = a \bullet b$ for the result that we wish to compute;

$\bar{a}, \bar{b}$ for representations of a and b;

$\bar{c} = \bar{a} \odot \bar{b}$ for the computed result;

$a \bullet b - \bar{a} \odot \bar{b}$ for the error in the elementary operation.

We introduce the following notation:

$$
\begin{aligned}
\delta(a) &= a - \bar{a} \\
\delta(b) &= b - \bar{b} \\
\delta(a \bullet b) &= a \bullet b - \bar{a} \odot \bar{b}
\end{aligned}
$$

The following results hold:

$$
\begin{aligned}
\delta(a + b) &= \delta(a) + \delta(b) \\
\delta(a - b) &= \delta(a) - \delta(b) \\
\frac{\delta(a \times b)}{a \times b} &= \frac{\delta(a)}{a} + \frac{\delta(b)}{b} \\
\frac{\delta(\frac{a}{b})}{\frac{a}{b}} &= \frac{\delta(a)}{a} - \frac{\delta(b)}{b}.
\end{aligned}
$$

REMARK.– Error propagates through any given sequence of elementary operations, since each intermediate calculation has its own errors.

2.5.2. *Operation of subtraction*

$$x \ominus y = \overline{\bar{x} - \bar{y}}$$

EXAMPLE.– Let $x = \sqrt{1001}$ and $y = \sqrt{1000}$.

In eight-digit floating-point arithmetic, we have

$$x \simeq 0.31638584 \times 10^2 \ \text{ and } y \simeq 0.31622777 \times 10^2.$$

Therefore, $x \ominus y = \overline{(0.31638584 - 0.31622777) \times 10^2} = 0.15807000 \times 10^{-1}$. We can obtain a more accurate result by using the following identity:

$$x - y = (\sqrt{1001} - \sqrt{1000}) \times \frac{\sqrt{1001} + \sqrt{1000}}{\sqrt{1001} + \sqrt{1000}} = \frac{1}{\sqrt{1001} + \sqrt{1000}}.$$

This yields:

$$1 \oslash (\sqrt{1001} \oplus \sqrt{1000}) = 1 \oslash (0.63261361 \times 10^2) = 0.15807437 \times 10^{-1}.$$

The true value is in fact $0.1580743742896 \times 10^{-1}$.

Subtraction is the most dangerous operation in numerical calculations. It can amplify the relative error catastrophically.

2.5.3. *Stability*

A calculation or an algorithm is said to be numerically stable if small variations in the data only generate small variations in the results. If this is not the case, the calculation or algorithm is said to be unstable.

2.6. Determining the best algorithm

Before we can identify the best possible algorithm, we need a way of comparing different algorithms. A good algorithm is an algorithm that is:

1) as inexpensive as possible in terms of memory usage;

2) as inexpensive as possible in terms of computation time, i.e. an algorithm that minimizes the number of required operations. This is known as a complexity problem;

EXAMPLE.– Suppose that we wish to compute the determinant of an $N \times N$ matrix. Cramer's method requires $N \times N!$ elementary operations ($N!$ additions and $(N-1)N!$ multiplications). If $N = 50$, noting that $50 \times 50! \simeq 1.52 \times 10^{66}$, and that a typical computer can perform roughly 20 GFlops (i.e. approximately 20 billion operations per second), we would need $1,050$ years to complete the calculation;

3) as stable as possible, i.e. as robust as possible with respect to the above-described rounding errors;

4) as accurate as possible, i.e. the solution obtained is an approximate solution, and we need to know how close it is to the exact solution. This process is called error estimation.

2.7. Using *Matlab*

2.7.1. *Definition of variables*

We do not need to declare variables and variable names, since *Matlab* does not distinguish between integer, real and complex variables.

In *Matlab*, i and j are reserved for the imaginary units ($i^2 = -1$ and $j^3 = -1$) by default. However, they can also be redefined as (integer or real) variables.

The following table presents the variable names and special values used by *Matlab*:

Variable names	Meaning	Value
eps	Relative precision of real numbers	2.2204×10^{-16}
pi	π	$3.14159\ldots$
i and j	Imaginary units	
inf	Infinity (1/0=inf)	∞
NAN	Not a number: $0/0 = NAN$	
flops	Counter of floating-point operations	

The *gamma* function returns the factorial of a number n. For example, for $n = 6$, we have: $6! = 6 * 5 * 4 * 3 * 2 * 1 = 720$.

```
>>factorial=gamma(6+1)

factorial =
                    720
```

The *gamma* function can of course calculate the factorial of integers, and also works with arbitrary real numbers.

2.7.2. *Manipulating numbers*

Matlab can use and manipulate complex numbers and variables. Most explicit functions defined for real numbers also exist for complex numbers, including power functions. For example:

```
>>z=3.5-1.25i ;
>>log(z)

ans =
    1.3128 - 0.3430i
>>cos(2+i)
```

```
ans =
     -0.6421 - 1.0686i
```

The pure imaginary part is indicated by i or j. Thus, typing i^2 or j^2 returns:

```
>>i^2

ans =
     -1
```

Let z be a complex number. The *conj* function returns its complex conjugate.

```
>>z=3.5-1.25i

z =
     3.5000 - 1.2500i
>>conj(z)

ans =
     3.5000 + 1.2500i
```

The following operations can be performed on complex numbers:

1) Addition of complex numbers:

```
>>z1

z1 =
     3.5000 - 1.2500i
>>z2

z2 =
     1.3140 - 0.0948i
>>z1+z2

ans =
     4.8140 - 1.3448i
```

2) Subtraction of complex numbers:

```
>>z1-z2
```

```
ans =
    2.186 - 1.1552i
```

3) Multiplication of complex numbers:

```
>>z1*z2

ans =
    4.4805 - 1.9743i
```

4) Division of complex numbers:

```
>>z1/z2

ans =
    2.7181 - 0.7551i
```

5) Power operation:

```
>>z1^z2

ans =
    4.5587 - 2.9557i
```

6) Modulus and argument of a complex number:

In *Matlab*, the functions *abs* and *angle* directly return the modulus and the argument of a complex number:

```
>>r=abs(z1)

r =
    3.7165
>>theta=angle(z1)

theta =
    -0.3430
```

The angle theta is in radians.

We can also use the exponential function:

```
>>Z=exp(z1)
Z =
    10.4420 - 31.4261i
```

2.7.2.1. *Numerical precision*

Matlab does not distinguish between integers, real numbers and complex numbers. Numbers are therefore handled automatically. The precision of a computation depends on the number of variables being stored and manipulated. The key parameters that determine this precision in any programming language are:

– the smallest positive number: x_min;

– the largest positive number: x_max;

– the machine error (epsilon): eps.

In *Matlab*, infinity is represented by “inf” (=). If we execute the command x=1/inf, it will return “NAN”, namely: “Not A Number”.

To find x_min, x_max and eps, we can run the following programs:

– finding x_min:

```
>>x=1;
while x>0,
x=x/2,
end
```

– finding x_max:

```
>>x=1;
while x<inf;
x= x*2;
end
```

– finding eps:

```
>>x=1;
while x>0;
x=x/2;
ex=0.98*x+1,
ex=ex-1,
if ex>0,ex;
end,
end
```

PART 2

Approximating Functions

3

Polynomial Interpolation

3.1. Introduction

Polynomial interpolation is a technique that interpolates a set of data points or a function using a polynomial. In other words, given a set of points, for example obtained from an experiment, we want to find the lowest-degree polynomial that passes through each point, possibly subject to additional constraints.

We will view polynomials as tools for solving interpolation problems, so the only algorithms that we will consider below are algorithms for evaluating polynomials or reformulating them with respect to different bases.

3.2. Interpolation problems

Interpolation problems are typically presented as follows:

A function f is known at the points $x_1, x_2, \ldots, x_n$, and nowhere else. This function f might have a complicated analytic expression, or the $f(x_i)$ could, for example, have been derived from physical measurements. For some x in the interval $[\min x_i, \max x_i]$, we now wish to evaluate $f(x)$ (when $x \notin [\min x_i, \max x_i]$, we say instead that the problem is an extrapolation problem).

An interpolation function or interpolant is a function p that satisfies:

$$p(x_i) = f(x_i) \quad i = 1, \ldots, n.$$

The process of solving the interpolation problem is generally divided into two smaller problems:

1) find an interpolation function (so that we can replace the computation of $f(x)$ by that of $p(x)$);

2) evaluate the interpolation error (when we replace $f(x)$ by $p(x)$, we are introducing an error $e(x) = f(x) - p(x)$ that we need to know how to estimate).

3.2.1. *Linear interpolation*

Linear interpolation works with the following framework. Let $p_1, p_2, \dots, p_n$ be n functions defined on the same interval I, generating a vector space $\mathcal{P}$. The interpolation problem may be stated as follows: given n pairs of real numbers:

$$(x_i, y_i) \quad i = 1, \dots, n \quad x_i \in I,$$

find $p \in \mathcal{P}$ such that:

$$p(x_i) = y_i \quad i = 1, \dots, n.$$

DEFINITION.– *The set $\{p_1, p_2, \dots, p_n\}$ is said to be unisolvent on the interval I if $\forall x_1, x_2, \dots, x_n \in I$ such that $x_i \neq x_j$ for $i \neq j$ (distinct points on the x-axis), the matrix G defined by:*

$$G = \begin{pmatrix} p_1(x_1) & p_2(x_1) & \cdots & p_n(x_1) \\ p_1(x_2) & p_2(x_2) & \cdots & p_n(x_2) \\ \vdots & \vdots & \ddots & \vdots \\ p_1(x_n) & p_2(x_n) & \cdots & p_n(x_n) \end{pmatrix}$$

is regular (this matrix G is called the Gram matrix).

THEOREM.– *The interpolation problem has a unique solution for any choice of pairs (x_i, y_i) (where $x_i \in I$ and $x_i \neq x_j$) if and only if the set $\{p_1, p_2, \dots, p_n\}$ is unisolvent on I.*

3.3. Polynomial interpolation techniques

This section studies the interpolation problem:

– find $p \in \mathcal{P}_n$ that interpolates the $(n+1)$ pairs (x_i, y_i) (where $i \neq j \rightarrow x_i \neq y_j$);

– find $p(x)$.

The interpolant p is fully determined by these $(n+1)$ pairs; finding p is therefore equivalent to calculating its coefficients with respect to some basis of $\mathcal{P}_n$. The properties of the algorithm for finding these coefficients will depend on the choice of basis. There are three common choices:

– the canonical basis;

– the Lagrange basis;

– the Newton basis.

3.4. Interpolation with the Lagrange basis

The Lagrange basis is especially suitable for interpolation problems. It is defined as a function of the $(n+1)$ distinct points x_i $(i = 0, \cdots, n)$.

We define the polynomials:

$$
\begin{aligned}
l_i(x) &= \frac{(x-x_0)(x-x_1)...(x-x_{i-1})(x-x_{i+1})...(x-x_n)}{(x_i-x_0)(x_i-x_1)...(x_i-x_{i-1})(x_i-x_{i+1})...(x_i-x_n)} \\
&= \prod_{\substack{j=0 \\ j \neq i}}^{n} \frac{x-x_j}{x_i-x_j}.
\end{aligned}
$$

Each polynomial l_i has degree n, and the following relations hold:

$$
l_i(x_j) = \delta_{ij} = \begin{cases} 1 \text{ if } i = j \\ 0 \text{ if } i \neq j \end{cases}.
$$

These $(n+1)$ polynomials are therefore linearly independent, and form a basis of $\mathcal{P}_n$.

It can immediately be seen that the polynomial:

$$
p(x) = \sum_{i=0}^{n} y_i l_i(x)
$$

satisfies the conditions $p(x_i) = y_i$ $(i = 0, \cdots, n)$ (the corresponding Gram matrix is the identity matrix). The Lagrange representation is therefore especially good for interpolation problems, since specifying the pairs (x_i, y_i) specifies the components y_i of the interpolating polynomial with respect to the Lagrange basis associated with the x_i.

However, to calculate $p(x)$, we need to calculate the $(n+1)$ values $l_i(x)$, which is expensive.

Total cost

$$\begin{array}{ll} (n+1)\times(2n-1) & \text{multiplicative operations} \\ +(n+1) & \text{multiplications } y_i l_i(x) \\ \hline 2n(n+1) & \text{multiplications} \end{array}$$

Another variant (described as the barycentric variant) of this algorithm is less expensive. We define:

$$\Pi(x) = \prod_{i=0}^{n}(x - x_i),$$

$$\Pi_i(x) = \prod_{\substack{j=0 \\ j \neq i}}^{n} (x - x_j).$$

Applying the identity $\sum_{i=0}^{n} l_i(x) = 1$ gives:

$$\Pi(x)\sum_{i=0}^{n}\frac{1}{(x - x_i)\Pi_i(x_i)} = 1.$$

By writing $B_i = \dfrac{1}{\Pi_i(x_i)}$,

$$\begin{aligned} \sum_{i=0}^{n} y_i l_i(x) &= \Pi(x)\sum_{i=0}^{n}\frac{y_i}{(x - x_i)\Pi_i(x_i)} \\ &= \Pi(x)\sum_{i=0}^{n} y_i\frac{B_i}{x - x_i} \\ &= \frac{\Pi(x)\sum_{i=0}^{n} y_i\dfrac{B_i}{x - x_i}}{\Pi(x)\sum_{i=0}^{n}\dfrac{B_i}{x - x_i}} \end{aligned}$$

$$= \frac{\sum_{i=0}^{n} y_i \frac{B_i}{x - x_i}}{\sum_{i=0}^{n} \frac{B_i}{x - x_i}}.$$

The cost of this method is:

$(n+1)^2$ for the $y_i \frac{B_i}{x - x_i}$

$n+1$ for the $\frac{B_i}{x - x_i}$

$(n+1)(n+2)$ multiplications.

Lagrange's method tends to be more interesting in theoretical contexts than in practice for the following reasons:

– high cost in terms of number of operations;

– awkward formulation. If we change any one of the points, we need to recalculate the Lagrange basis. Similarly, if we want to add another pair (x_{n+1}, y_{n+1}), we have to start from scratch;

– on the contrary, we can replace the y_i without the need to recalculate the basis.

EXAMPLE.– Suppose that the three points $(0, 1)$, $(2, 5)$ and $(4, 17)$ are known. The corresponding Lagrange polynomials are:

$$L_0(x) = \frac{(x-2)(x-4)}{8}, \quad L_1(x) = \frac{x(x-4)}{-4}, \quad L_2(x) = \frac{x(x-2)}{8}.$$

The interpolating polynomial is:

$$p(x) = L_0(x) + 5L_1(x) + 17L_2(x) = x^2 + 1.$$

EXAMPLE.–

1) Suppose that we wish to find the Lagrange interpolating polynomial of the following array of points:

k	1	2	3
x_k	-1	0	1
y_k	1	3	2

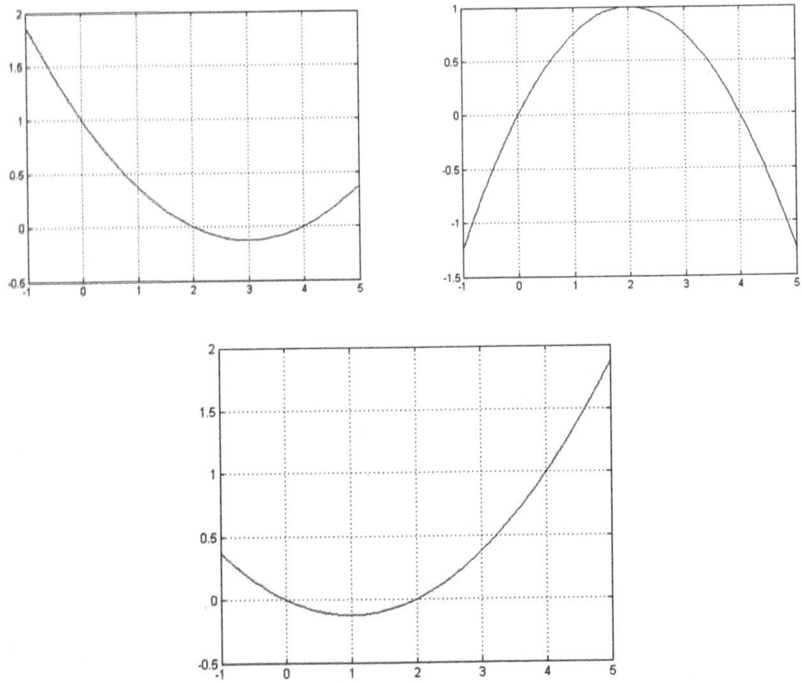

Figure 3.1. *Three examples of Lagrange functions*

This interpolating polynomial $P(x)$ satisfies $P(x_k) = y_k$ by Lagrange's method, so:

$$P(x) = L_1(x)y_1 + L_2(x)y_2 + L_3(x)y_3.$$

The Lagrange polynomials $L_i(x)$ are:

$$L_1(x) = \frac{(x-x_2)(x-x_3)}{(x_1-x_2)(x_1-x_3)} = \frac{x(x-1)}{2},$$

$$L_2(x) = \frac{(x-x_1)(x-x_3)}{(x_2-x_1)(x_2-x_3)} = -(x^2-1),$$

$$L_3(x) = \frac{(x-x_1)(x-x_2)}{(x_3-x_1)(x_3-x_2)} = \frac{x(x+1)}{2}.$$

Hence:

$$\begin{aligned} P(x) &= \frac{x(x-1)}{2} \times 1 - (x^2-1) \times 3 + \frac{x(x+1)}{2} \times 2 \\ &= -\frac{3}{2}x^2 + \frac{1}{2}x + 3. \end{aligned}$$

2) Suppose instead that we wish to find the interpolating polynomial directly. Setting $P(x) = ax^2 + bx + c$ with $P(x_k) = y_k$ yields the following system:

$$\begin{cases} ax_1^2 + bx_1 + c = y_1 \\ ax_2^2 + bx_2 + c = y_2 \\ ax_3^2 + bx_3 + c = y_3 \end{cases} \Rightarrow \begin{cases} a - b + c = 1 \\ c = 3 \\ a + b + c = 2 \end{cases}$$

Solving this system gives $a = -\frac{3}{2}$, $b = \frac{1}{2}$ and $c = 3$. Therefore:

$$P(x) = -\frac{3}{2}x^2 + \frac{1}{2}x + 3.$$

3.4.1. *Polynomial interpolation error*

Let p be a polynomial with degree less than or equal to n that interpolates a function f at the $(n+1)$ points x_i. The value $e(x) = f(x) - p(x)$ is called the *interpolation error*.

Let I be the interval:

$$I = \left[\min_i\{(x_i), x\}, \max_i\{(x_i), x\}\right].$$

THEOREM.– *If f is $(n+1)$ times continuously differentiable on I, then:*

$$\exists \xi \in I \;:\; e(x) = (x - x_0)(x - x_1)\ldots(x - x_n)\frac{f^{(n+1)}(\xi)}{(n+1)!}.$$

EXAMPLE.– Suppose that the three points $(0, 1)$, $(2, 7.3891)$ and $(4, 54.5982)$ are known. The interpolating polynomial is:

$$p(x) = L_0(x) + 7.3891L_1(x) + 54.5982L_2(x).$$

In fact, the function that we are interpolating is $\exp x$. The curves of the two functions are shown in Figure 3.2, and Figure 3.3 shows the interpolation error.

THEOREM.– *Suppose that f is $(n+1)$ times continuously differentiable on the interval $[a, b]$. Then, by writing p_n for the interpolating polynomial of degree n, the following relation holds:*

$$\max_{x\in[a,b]} |f(x) - p_n(x)| \leq \frac{1}{2(n+1)}\left(\frac{b-a}{n}\right)^{(n+1)} \max_{x\in[a,b]} |f^{(n+1)}(x)|, \qquad [3.1]$$

where $f^{(n+1)}(x) = d^{(n+1)}f(x)/dx^{n+1}$.

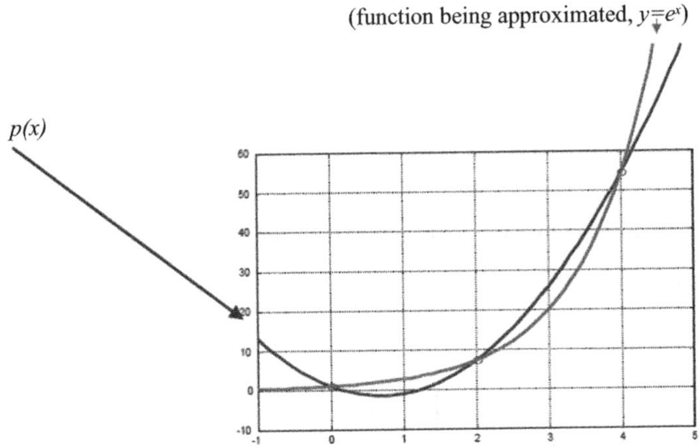

Figure 3.2. *The function* $\exp x$ *and its interpolant* $p(x)$

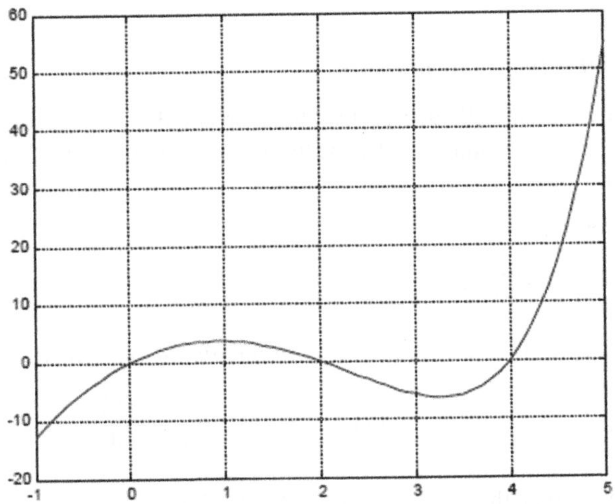

Figure 3.3. *Interpolation error relative to the function* $\exp x$

The inequality in [3.1] estimates the error between the function f and its interpolant p_n of degree n for the points $x_0, x_1, \ldots, x_n$, which are equally spaced in $[a, b]$. *A priori*, we might expect this error to tend to zero as n tends to infinity, since:

$$\lim_{n \to +\infty} \frac{1}{2(n+1)} \left(\frac{b-a}{n} \right)^{(n+1)} = 0.$$

However, this claim often fails, since $\max_{x\in[a,b]} |f^{(n+1)}(x)|$ can increase very rapidly with n. This is known as Runge's phenomenon.

EXAMPLE.– Consider the function $f(x) = \sin \frac{\pi}{2}x$.

Suppose that $p(x)$ is the quadratic interpolating polynomial of f at the points $x_i = 0, 1, 2$. We know that $f(0) = 0$, $f(1) = 1$ and $f(2) = 0$. We need to find $p(x)$ such that $p(x) = a_0 + a_1x + a_2x^2$, with $p(0) = 0$, $p(1) = 1$ and $p(2) = 0$. Thus:

$$\begin{cases} a_1 + a_2 &= 1 \\ 2a_1 + 4a_2 &= 0 \end{cases} \Leftrightarrow \begin{cases} a_2 = -1 \\ a_1 = \ \ 2 \end{cases}$$

and so:

$$p(x) = -x^2 + 2x.$$

We can now find an expression for the error $f(x)-p(x)$. This error may be written in the form:

$$E(x) = f(x) - p(x) = \frac{f^{(3)}(\xi_x}{3!}\pi(x),$$

where $\pi(x) = x(x-1)(x-2)$ and $\xi_x \in [0, 2]$.

We can show that:

$$|E(x)| = |f(x) - p(x)| \leq \left|\frac{\pi^3}{48}x(x-1)(x-2)\right|.$$

Indeed, $f^{(3)}(x) = -\frac{\pi^3}{8}\cos\frac{\pi}{2}x$, so:

$$\begin{aligned} |f(x) - p(x)| &\leq \frac{\pi^3}{48} \sup_{x\in[0,2]} |\cos\frac{\pi}{2}x||\pi(x)| \\ &\leq \left|\frac{\pi^3}{48}x(x-1)(x-2)\right|. \end{aligned}$$

Our estimate of the error for $x = \frac{1}{2}$ is $\frac{\pi^3}{128}$, which is approximately:

$$\frac{\pi^3}{128} \simeq 0.242 = 24.2 \times 10^{-2}. \qquad [3.2]$$

We can find the true error as follows:

$$f\left(\frac{1}{2}\right) = \sin\frac{\pi}{4} = \frac{\sqrt{2}}{2} \text{ and } p\left(\frac{1}{2}\right) = \frac{3}{4},$$

so:

$$\left|f(\frac{1}{2}) - p(\frac{1}{2})\right| = \left|\frac{\sqrt{2}}{2} - \frac{3}{4}\right| \simeq 4.2 \times 10^{-2}.$$

Therefore, the error estimate given by [3.2] is less tight than the true error.

3.4.2. *Neville–Aitken method*

The Neville–Aitken method is a recursive algorithm for calculating the Lagrange interpolating polynomial on n points from the expression on $(n-1)$ points. To calculate an interpolant of f at a point x, we can use the divided difference formulas. However, the method proposed by Aitken avoids the need to compute the coefficients of the polynomial, and does not require us to assume that the points x_i are equally spaced. The algorithm is based on the following result:

Let $f(x|x_p,\ldots,x_p)$ be the unique interpolating polynomial of degree $(q-p-1)$ which is equal to $f(x)$ at the points $x_p,\ldots,x_q$. The following recurrence relation holds:

$$(x_k-x_j)f(x|x_p,\ldots,x_q) = \begin{vmatrix} (x_k-x) & f(x|x_p,\ldots,x_{j-1},x_{j+1},\ldots,x_q) \\ (x_j-x) & f(x|x_p,\ldots,x_{k-1},x_{k+1},\ldots,x_q) \end{vmatrix}. \quad [3.3]$$

In practice, we write the computation to find the interpolating polynomial in the form of a table. Let $P_{i+1,j}$ be the Lagrange interpolating polynomial of the points $x_j,x_{j+1},\ldots,x_{j+i+1}$. The above formula may be written as:

$$(x_{i+j+1}-x_j)P_{i+1,j} = (x_{i+j+1}-x)P_{i,j} - (x_j-x)P_{i,j+1}.$$

Suppose that the values $f(x_i) = f_i = P_{0,i}$ are known for $i = 0,1,2,3$:

$$\begin{array}{l} f_0 = P_{0,0}\ P_{1,0}\ P_{2,0}\ P_{3,0} \\ f_1 = P_{0,1}\ P_{1,1}\ P_{2,1} \\ f_2 = P_{0,2}\ P_{1,2} \\ f_3 = P_{0,3} \end{array}$$

We can fill out each column of the table in turn using this formula. The value $P_{3,0}$ is the desired result.

3.5. Interpolation with the Newton basis

Let p_k be the polynomial that interpolates the function f at the points $x_0,x_1,\ldots,x_k$ for $k = 0,1,\ldots,n$.

We will construct p_n inductively.

– In the initial case, $p_0(x) = f(x_0)$ (polynomial of degree 0).

– For $k \geq 1$, the polynomial $p_k - p_{k-1}$ has degree k and vanishes at the points $x_0,\ x_1,\ldots,\ x_{k-1}$. It is therefore of the form:

$$p_k(x) - p_{k-1}(x) = f[x_0,x_1,\cdots,x_k](x-x_0)(x-x_1)\cdots(x-x_{k-1}),$$

where $f[x_0,x_1,\cdots,x_k]$ denotes the coefficient of x^k in $p_k(x)$.

Summing these relations for $k = 1, 2, \cdots, n$:

$$p_n(x) = f(x_0) + \sum_{k=1}^{n} f[x_0, x_1, \cdots, x_k](x - x_0)(x - x_1) \cdots (x - x_{k-1}). \text{[3.4]}$$

REMARK.– The set of $n + 1$ polynomials:

$$\{1; (x - x_0); (x - x_0)(x - x_1); \cdots; (x - x_0)(x - x_1) \cdots (x - x_{n-1})\}$$

forms a basis of the space $\mathcal{P}_n$, called the *Newton basis.*

Formula [3.4] reduces the recursive computation of p_n to the computation of the coefficients $f[x_0, x_1, \ldots, x_k]$.

DEFINITION.– *The quantity $f[x_0, x_1, \ldots, x_k]$ is called the k-th divided difference of f for the points $x_0, x_1, \ldots, x_k$.*

LEMMA.–

$$f[x_i] = f(x_i), \; i = 0, \ldots, k;$$

$$\forall k \geq 1, \; f[x_0, x_1, \ldots, x_k] = \frac{f[x_1, \ldots, x_k] - f[x_0, x_1, \ldots, x_{k-1}]}{x_k - x_0}.$$

LEMMA.– *The Newton interpolation error is*

$$e_n(x) = f(x) - p_n(x) = \Pi(x).f[x_0, x_1, \ldots, x_n, x].$$

If $f \in C^{n+1}([a, b])$, then $\exists \xi \in [a, b]$ such that

$$f[x_0, x_1, \ldots, x_n, x] = \frac{f^{(n+1)}(\xi)}{(n+1)!}.$$

REMARK.– By definition, the k-th divided difference of f for the points x_0, x_1,..., x_n, $f[x_0, x_1, \ldots, x_k]$ is independent of the order of the points x_i.

For example, we will calculate the divided differences of data points in the following table:

k	1	2	3	4
x_1	0	1	2	4
y_k	1	1	2	5

The divided differences are:

$$
\begin{array}{ccccc}
x_k & y_k & & & \\
0 & 1 & & & \\
 & & 0 & & \\
1 & 1 & & \frac{1}{2} & \\
 & & 1 & & -\frac{1}{12} \\
2 & 2 & & \frac{1}{6} & \\
 & & \frac{3}{2} & & \\
4 & 5 & & &
\end{array}
$$

The Newton interpolating polynomial of this table is:

$$
\begin{aligned}
P(x) &= y_0 + (x - x_0)y(x_0, x_1) + (x - x_0)(x - x_1)y(x_0, x_1, x_2) \\
&\quad + (x - x_0(x - x_1)(x - x_2)y(x_0, x_1, x_2, x_3)
\end{aligned} \qquad [3.5]
$$

so:

$$
\begin{aligned}
P(x) &= 1 + (x - 0) \times 0 + (x - 0)(x - 1) \times \frac{1}{2} \\
&\quad + (x - 0)(x - 1)(x - 2) \times -\frac{1}{12} \\
&= 1 - \frac{2}{3}x + \frac{3}{4}x^2 - \frac{x^3}{12}.
\end{aligned}
$$

3.6. Interpolation using spline functions

The word "spline" literally means a "flexible strip". The idea is to study the curve of a strip that is forced to pass through finitely many given points (namely, (x_i, y_i) for $i = 0, 1, \dots, n$). Figure 3.4 shows the spline that passes through the same points as in Figure 3.5 (in comparison with the interpolating polynomial).

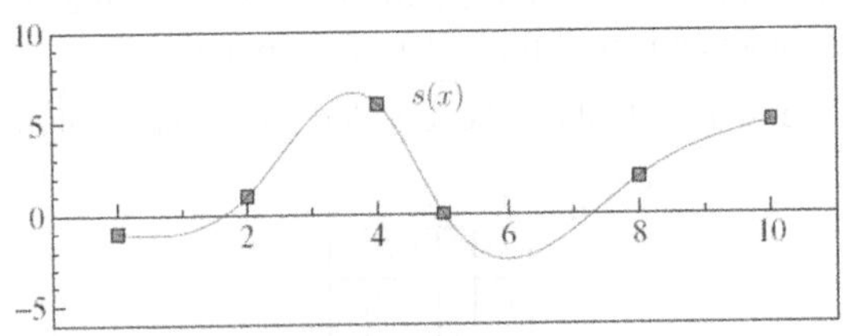

Figure 3.4. *Cubic spline*

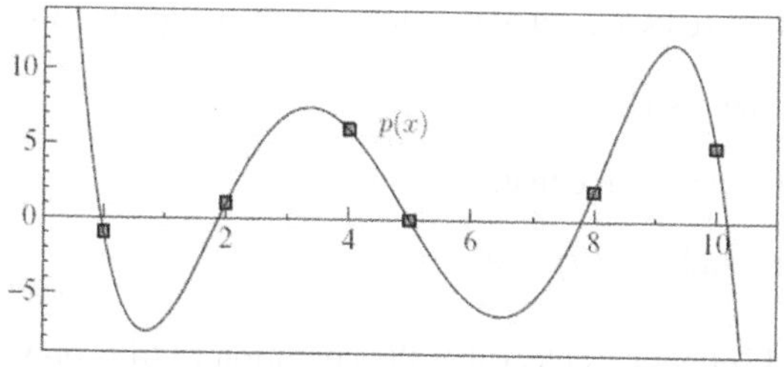

Figure 3.5. *Newton interpolation*

For a given set of points (x_i, y_i), we want to find a function $s\colon [a,b] \to \mathbb{R}$ (where $a = x_0$, $b = x_n$) satisfying:

(S1) $s(x_i) = y_i$ for $i = 0, 1, \ldots, n$;

(S2) $s(x)$ is twice continuously differentiable (i.e. has class C^2);

(S3) $\int_a^b (s''(x))^2 dx$ is minimal.

The integral of (S3) represents the energy of the deformed strip. By Maupertuis' principle, we assume that this energy is minimal.

THEOREM.– *Let $a = x_0 < x_1 < \ldots < x_n = b$ be a partition, and suppose that $s\colon [a,b] \to \mathbb{R}$ is a function satisfying (S1) and (S2). Suppose further that the function s is a cubic polynomial on each interval $[x_{i-1}, x_i]$.*

Then, given any function $f\colon [a,b] \to \mathbb{R}$ that satisfies (S1), (S2) and:

$$s''(b)\,(f'(b) - s'(b)) = s''(a)\,(f'(a) - s'(a)), \qquad [3.6]$$

we have that:

$$\int_a^b (s''(x))^2 dx \leq \int_a^b (f''(x))^2 dx. \qquad [3.7]$$

The above theorem shows that one class of candidates that satisfies (S1)–(S3) is given by the piecewise cubic polynomials with the continuity class C^2.

DEFINITION.– *Let $a = x_0 < x_1 < \ldots < x_n = b$ be a partition of $[a,b]$. A function $s\colon [a,b] \to \mathbb{R}$ is said to be a (cubic) spline if it is twice continuously differentiable and is a cubic polynomial on each interval $[x_{i-1}, x_i]$.*

To satisfy the condition [3.6], we have several options:

– Natural spline: we assume that:

$$s''(a) = 0 \quad \text{and} \quad s''(b) = 0. \qquad [3.8]$$

– Clamped spline: we assume that the slopes at the endpoints are given, i.e.:

$$s'(a) = p_0 \quad \text{and} \quad s'(b) = p_n. \tag{3.9}$$

– Periodic spline: we assume that:

$$s'(a) = s'(b) \quad \text{and} \quad s''(a) = s''(b). \tag{3.10}$$

For the clamped spline, it is clear that the condition [3.6] is only satisfied if $f(x)$ also satisfies the condition [3.9]. If so, $s(x)$ only minimizes the integral of (S3) over the class of functions with fixed slopes at the endpoints. Similar remarks hold for the periodic spline.

3.6.1. *Hermite interpolation*

The objective of this section is to derive a construction of the spline that satisfies $s(x_i) = y_i$ for $i = 0, 1, \ldots, n$, as well as one of the conditions [3.8–3.10].

Consider one of the intervals $[x_{i-1}, x_i]$. We will look for a cubic polynomial $s_i(x)$ satisfying:

$$\begin{aligned} &s_i(x_{i-1}) = y_{i-1},\ s_i(x_i) = y_i \\ &s_i'(x_{i-1}) = p_{i-1},\ s_i'(x_i) = p_i. \end{aligned} \tag{3.11}$$

The solution can be obtained from Newton's formula, replacing the last two conditions in [3.11] by $s_i(x_{i-1} + \epsilon) = y_{i-1} + \epsilon p_{i-1}$ and $s_i(x_i - \epsilon) = y_i - \epsilon p_i$, and evaluating the limit as $\epsilon \to 0$. The divided differences corresponding to the points in [3.11] are listed in Table 3.1 ($h_{i-1} = x_i - x_{i-1}$). By substituting the values in this table into Newton's formula, we find that:

$$\begin{aligned} s_i(x) &= y_{i-1} + (x - x_{i-1})\delta y[x_i, x_{i-1}] \\ &+ \frac{(x - x_{i-1})(x - x_i)}{h_{i-1}^2}\left((p_i - \delta y[x_i, x_{i-1}])(x - x_{i-1})\right. \\ &\left.+ (p_{i-1} - \delta y[x_i, x_{i-1}])(x - x_i)\right). \end{aligned} \tag{3.12}$$

Construction of the interpolating spline: for each choice of slopes p_0, p_1,..., p_n, the function $s\colon [a, b] \to \mathbb{R}$ defined by $s(x) = s_i(x)$ for $x \in [x_{i-1}, x_i]$ satisfies the following properties:

a) $s(x_i) = y_i$ for $i = 0, 1, \ldots, n$;

b) $s(x)$ has class C^1 and $s'(x_i) = p_i$ for $i = 0, 1, \ldots, n$;

c) on each interval $[x_{i-1}, x_i]$, $s(x)$ is a cubic polynomial.

x_{i-1}	y_{i-1}			
		p_{i-1}		
x_{i-1}	y_{i-1}		$h_{i-1}^{-1}(\delta y[x_i, x_{i-1}] - p_{i-1})$	
		$\delta y[x_i, x_{i-1}]$		$h_{i-1}^{-2}(p_i + p_{i-1} - 2\delta y[x_i, x_{i-1}])$
x_i	y_i		$h_{i-1}^{-1}(p_i - \delta y[x_i, x_{i-1}])$	
		p_i		
x_i	y_i			

Table 3.1. *Divided differences for Hermite interpolation*

To construct the interpolating spline, the slopes p_0, p_1,..., p_n must be chosen in such a way that s'' is continuous, i.e.:

$$s_i''(x_i) = s_{i+1}''(x_i) \quad i = 1, \ldots, n-1, \tag{3.13}$$

and one of the conditions [3.8–3.10] is satisfied. By differentiating [3.12] twice, we find:

$$s_i''(x_i) = \frac{2}{h_{i-1}}(2p_i + p_{i-1} - 3\delta y[x_i, x_{i-1}]),$$
$$s_i''(x_{i-1}) = -\frac{2}{h_{i-1}}(p_i + 2p_{i-1} - 3\delta y[x_i, x_{i-1}]).$$

The condition [3.13] thus becomes (for $i = 1, \ldots, n-1$):

$$\frac{p_{i-1}}{h_{i-1}} + 2p_i\left(\frac{1}{h_{i-1}} + \frac{1}{h_i}\right) + \frac{p_{i+1}}{h_i} = 3\left(\frac{\delta y[x_i, x_{i-1}]}{h_{i-1}} + \frac{\delta y[x_{i+1}, x_i]}{h_i}\right). \tag{3.14}$$

This gives $n-1$ linear equations in the $n+1$ unknowns $p_0, p_1, \ldots, p_n$. The two final conditions depend on the type of spline. For example, in the case of the clamped spline, the values of p_0 and p_n are explicitly specified, and the resulting linear system may be written in matrix form as follows:

$$\underbrace{\begin{pmatrix} 2\left(\frac{1}{h_0}+\frac{1}{h_1}\right) & \frac{1}{h_1} & & & \\ \frac{1}{h_1} & 2\left(\frac{1}{h_1}+\frac{1}{h_2}\right) & \frac{1}{h_2} & & \\ & \frac{1}{h_2} & 2\left(\frac{1}{h_2}+\frac{1}{h_3}\right) & \ddots & \\ & & \ddots & \ddots & \frac{1}{h_{n-2}} \\ & & & \frac{1}{h_{n-2}} & 2\left(\frac{1}{h_{n-2}}+\frac{1}{h_{n-1}}\right) \end{pmatrix}}_{A} \begin{pmatrix} p_1 \\ p_2 \\ p_3 \\ \vdots \\ p_{n-1} \end{pmatrix}$$

$$= \begin{pmatrix} c_1 \\ c_2 \\ c_3 \\ \vdots \\ c_{n-1} \end{pmatrix}, \qquad [3.15]$$

where the c_i are given by [3.14]. The matrix A is symmetric and tridiagonal. It can be shown that A is invertible. Therefore, the clamped spline always exists and is unique. The same approach works for the natural and periodic splines.

We can solve the above linear system by elimination, first eliminating the variable p_1 in the second row using the first row, then the variable p_2 in the third row using the second row, and so on. This yields a bidiagonal system that is easy to solve.

EXAMPLE.– Let f be a function with class C^1 on $[a, b]$. We wish to find a polynomial p of degree at most 3 such that:

$$\begin{aligned} p(a) &= f(a); \\ p(b) &= f(b); \\ p'(a) &= f'(a); \\ p'(b) &= f'(b). \end{aligned} \qquad [3.16]$$

The function f has class C^1 on $[a, b]$. We will look for a cubic polynomial that satisfies the equations in [3.16].

After setting $p(x) = a_0 + a_1 x + a_2 x^2 + a_3 x^3$, the system in [3.16] is satisfied if and only if:

$$\begin{cases} a_0 + a_1 a + a_2 a^2 + a_3 a^3 = f(a) \\ a_0 + a_1 b + a_2 b^2 + a_3 b^3 = f(b) \\ a_1 + 2a_2 a + 3a_3 a^2 = f'(a) \\ a_1 + 2a_2 b + 3a_3 b^2 = f'(b). \end{cases}$$

In matrix form:

$$\underbrace{\begin{pmatrix} 1 & a & a^2 & a^3 \\ 1 & b & b^2 & b^3 \\ 0 & 1 & 2a & 3a^2 \\ 0 & 1 & 2b & 3b^2 \end{pmatrix}}_{M} \begin{pmatrix} a_0 \\ a_1 \\ a_2 \\ a_3 \end{pmatrix} = \begin{pmatrix} f(a) \\ f(b) \\ f'(a) \\ f'(b) \end{pmatrix}.$$

The polynomial p exists and is unique if and only if $\det M \neq 0$ (i.e. M is an invertible matrix). Now:

$$\begin{aligned}
\det M &= \begin{vmatrix} 1 & a & a^2 & a^3 \\ 0 & b-a & b^2-a^2 & b^3-a^3 \\ 0 & 1 & 2a & 3a^2 \\ 0 & 0 & 2(b-a) & 3(b^2-a^2) \end{vmatrix} \\
&= \begin{vmatrix} b-a & b^2-a^2 & b^3-a^3 \\ 1 & 2a & 3a^2 \\ 0 & 2(b-a) & 3(b^2-a^2) \end{vmatrix} \\
&= (b-a)^2 \begin{vmatrix} 1 & b+a & b^2+ab+a^2 \\ 1 & 2a & 3a^2 \\ 0 & 2 & 3(b+a) \end{vmatrix} \\
&= (b-a)^2 \begin{vmatrix} 1 & b+a & a^2+ab+b^2 \\ 0 & a-b & 2a^2-ab-b^2 \\ 0 & 2 & 3(b+a) \end{vmatrix} \\
&= -(b-a)^4.
\end{aligned}$$

Moreover, $b \neq a$, so $\det M \neq 0$, and therefore there exists precisely one cubic polynomial that satisfies the system of equations in [3.16].

Let φ_a and φ_b be the Lagrange basis vectors corresponding to the points a and b.

We will now find the polynomials Q_a, Q_b, R_a and R_b such that:

$$\begin{aligned}
p(x) &= Q_a(x)\,\varphi_a^2(x)\,f(a) + Q_b(x)\,\varphi_b^2(x)\,f(b) \\
&\quad + R_a(x)\,\varphi_a^2(x)\,f'(a) + R_b(x)\,\varphi_b^2(x)\,f'(b).
\end{aligned}$$

We have that:

$$\phi_a(x) = \frac{x-b}{a-b} \quad \phi_b(x) = \frac{x-a}{b-a}.$$

The polynomials Q_a, Q_b, R_a and R_b will each have degree 1. To find polynomials satisfying:

$$p(x) = Q_a(x)\varphi_a^2(x) f(a) + Q_b(x)\varphi_b^2(x) f(b) + R_a(x)\varphi_a^2(x) f'(a) + R_b(x)\varphi_b^2(x) f'(b),$$

we can rewrite the conditions in [3.16] and choose sufficient conditions that allow us to deduce the desired polynomials by identification. For example, for the polynomial:

$$p(a) = Q_a(a)f(a) + R_a(a)f'(a) = f(a),$$

we can choose $\phi_a(a) = 1$ and $R_a(a) = 0$. For:

$$p(b) = Q_b(b)f(b) + R_b(b)f'(b) = f(b),$$

we can choose $\phi_b(b) = 1$ and $R_b(b) = 0$.

The expression of $p'(x)$ is:

$$\begin{aligned} p'(x) &= [Q_a'(x)\phi_a^2(x) + 2\phi_a(x)\phi_a'(x)Q_a(x)]f(a) \\ &+ [Q_b'(x)\phi_b^2(x) + 2\phi_b(x)\phi_b'(x)Q_b(x)]f(b) \\ &+ [R_a'(x)\phi_a^2(x) + 2R_a(x)\phi_a(x)\phi_a'(x)]f'(a) \\ &+ [R_b'(x)\phi_b^2(x) + 2R_b(x)\phi_b(x)\phi_b'(x)]f'(b), \end{aligned}$$

so:

$$p'(a) = [Q_a'(a) + 2\phi_a'(a)]f(a) + R_a'(a) \times f'(a).$$

We can choose $R_a'(a) = 1$ and $Q_a'(a) + 2Q_a'(a) = 0$, so:

$$R_a(x) = x - a \quad Q_a(x) = \frac{2x + b - 3a}{b - a}.$$

Proceeding similarly for b, we find:

$$R_b(x) = x - b \quad Q_b(x) = \frac{2x + a - 3b}{a - b}.$$

Suppose that $f \in C^4([a,b])$. It can be shown that $\forall x \in [a,b], \exists \lambda_x \in]a,b[$ such that:

$$f(x) = p(x) + \pi^2(x)\frac{f^{(4)}(\lambda_x)}{4!}, \text{ where } \pi(x) = (x-a)(x-b).$$

This equality also holds at $x = a$ and $x = b$. Now, consider:

$$F(t) = f(t) - p(t) - \left(\frac{f(x) - p(x)}{\pi^2(x)}\right)\pi^2(t).$$

Since f has class C^4, so does F. It can easily be verified that x, a and b are roots of the equation $F(t) = 0$. Therefore, this equation has at least three roots in $[a, b]$.

By Rolle's theorem, $F'(t) = 0$ has at least two roots that are distinct from x, a and b. It can also be verified that a and b are roots of $F'(t) = 0$. Therefore, F' has at least four roots in $[a, b]$, and so $F^{(4)}$ has at least one root between a and b. This implies that there exists $\lambda_x \in I$ such that:

$$F^{(4)}(\lambda_x) = 0 = f^{(4)}(\lambda_x) - \underbrace{p^{(4)}(\lambda_x)}_{=0} - \frac{f(x) - p(x)}{\pi^2(x)} 4!,$$

and therefore:

$$f(x) = p(x) + \pi^2(x) \frac{f^{(4)}}{4!}(\lambda_x). \qquad [3.17]$$

3.6.2. *Spline interpolation error*

Consider a differentiable function $f\colon [a, b] \to \mathbb{R}$ and a partition $a = x_0 < x_1 < \ldots < x_n = b$. Suppose that the (clamped) spline $s(x)$ satisfies:

$$s(x_i) = f(x_i), \ \ i = 0, 1, \ldots, n, \qquad [3.18]$$

$$s'(x_0) = f'(x_0), \ s'(x_n) = f'(x_n). \qquad [3.19]$$

One expression for this spline is given by [3.12], where the coefficients p_i satisfy [3.14], $p_0 = f'(x_0)$ and $p_n = f'(x_n)$. Our next task is to study the error $f(x) - s(x)$ for $x \in [a, b]$.

To do this, the key idea is to consider the Hermite interpolating polynomial (on $[x_{i-1}, x_i]$) $q_i(x)$, which is defined by:

$$\begin{aligned} &q_i(x_{i-1}) = f(x_{i-1}), \ \ q_i(x_i) = f(x_i), \\ &q_i'(x_{i-1}) = f'(x_{i-1}), \ \ q_i'(x_i) = f'(x_i). \end{aligned} \qquad [3.20]$$

An expression for this polynomial is also given by [3.12], replacing the terms p_j in the formula by $f'(x_j)$. We will estimate the two terms on the right-hand side of

$$f(x) - s_i(x) = (f(x) - q_i(x)) + (q_i(x) - s_i(x)) \qquad [3.21]$$

separately.

THEOREM.– *Suppose that $f(x)$ has class C^4 on $[a, b]$ and let $q_i(x)$ be the cubic polynomial that satisfies [3.20] (the Hermite interpolating polynomial). Then, for $x \in [x_{i-1}, x_i]$,*

$$|f(x) - q_i(x)| \leq \frac{h_{i-1}^4}{384} \times \max_{\xi \in [x_{i-1}, x_i]} |f^{(4)}(\xi)|. \qquad [3.22]$$

To estimate the second term in [3.21], we can subtract the formula in [3.12] for $s_i(x)$ from the corresponding formula for $q_i(x)$. This gives:

$$q_i(x) - s_i(x) = \frac{(x - x_{i-1})(x - x_i)}{h_{i-1}^2}\left((f'(x_i) - p_i)(x - x_{i-1})\right.$$
$$\left. + (f'(x_{i-1}) - p_{i-1})(x - x_i)\right). \qquad [3.23]$$

We still need to estimate $f'(x_i) - p_i$.

LEMMA.– *Let $f(x)$ have class C^4 on $[a, b]$ and write $h = \max_i h_i$, where $h_i = x_{i+1} - x_i$. If the p_i satisfy [3.14], $p_0 = f'(x_0)$ and $p_n = f'(x_n)$, then:*

$$|f'(x_i) - p_i| \leq \frac{h^3}{24} \times \max_{x \in [a,b]} |f^{(4)}(x)|. \qquad [3.24]$$

If the partition is equally spaced and $f \in C^5([a, b])$, then:

$$|f'(x_i) - p_i| \leq \frac{h^4}{60} \times \max_{x \in [a,b]} |f^{(5)}(x)|. \qquad [3.25]$$

This implies the following theorem:

THEOREM.– *Suppose that $f\colon [a, b] \to \mathbb{R}$ has class C^4, let $a = x_0 < x_1 < \ldots < x_n = b$ be an arbitrary partition, and let $s(x)$ be the spline passing through $(x_i, f(x_i))$ for $i = 0, 1, \ldots, n$ that satisfies $s'(x_0) = f'(x_0)$ and $s'(x_n) = f'(x_n)$. Then, by writing $h_i = x_{i+1} - x_i$ and $h = \max_i h_i$,*

$$|f(x) - s(x)| \leq \frac{5}{384} h^4 \times \max_{\xi \in [a,b]} |f^{(4)}(\xi)|. \qquad [3.26]$$

In particular, if the partition is equally spaced and $f \in C^5[a, b]$, then:

$$|f(x) - s(x)| \leq \frac{h^4}{384} \times \max_{\xi \in [a,b]} |f^{(4)}(\xi)| + \frac{h^5}{240} \times \max_{\xi \in [a,b]} |f^{(5)}(\xi)|. \qquad [3.27]$$

EXAMPLE.– Consider the real function $f(x) = \dfrac{1}{1 + x^2}$.

We will use this example to compare the performance of each interpolation method by calculating $f(4)$ (whose exact value is known).

Let P be the Lagrange interpolating polynomial of the function f at the points $0, 1, 3, 5$. We will calculate $P(4)$ and find an upper bound for the error $|f(4) - P(4)|$ by calculating $f^{(4)}(x)$:

$$P(x) = -\frac{3}{260}x^3 + \frac{19}{130}x^2 - \frac{33}{52}x + 1$$

$$P(4) = \frac{4}{65} \simeq 0.06153846154$$

$$f^{(4)}(x) = 24\frac{5x^4 - 10x^2 + 1}{(x^2+1)^5}$$

$|f(4) - P(4)| \leq \dfrac{|\Pi(4)|}{4!} \sup_{t\in[0,5]} |f^{(4)}(t)|$. We know that $f(4) = \dfrac{1}{17}$, so the exact error is $f(4) - P(4) = \dfrac{1}{17} - \dfrac{4}{65} = -\dfrac{3}{1105} \simeq -3 \times 10^{-3}$.

Next, we will find the Hermite polynomial Q such that:

$$Q(0) = f(0) \quad Q'(0) = f'(0) \quad Q(5) = f(5) \quad Q'(5) = f'(5)$$

$$Q(x) = \frac{(x-5)^2}{25} + \frac{2x(x-5)^2}{125} + \frac{1}{26}(\frac{x^2}{25} - \frac{2x^2(x-5)}{125}) - \frac{10}{26^2}\frac{x^2(x-5)}{25}.$$

Now, we compute $Q(4)$ and find an upper bound for the error $|f(4) - Q(4)|$.

$$Q(4) = \frac{25}{169} \simeq 0.1479289941.$$

An upper bound for the error is:

$$|f(x) - Q(x)| \leq \frac{(x-0)^2(x-5)^2}{24} M_4.$$

Here, $M_4 = 24$ and $|f(4) - Q(4)| \leq 16$, so the true error is:

$$f(4) - Q(4) = \frac{1}{17} - \frac{25}{169} \simeq -0.08910546467.$$

Finally, we will find the cubic spline R interpolating f at the points $1, 4, 5$ using both methods (first with the condition $R''(1) = R'(5) = 0$, then with the conditions $R'(1) = f'(1)$ and $R'(5) = f'(5)$).

We need to find $(\lambda_0, \lambda_1, \lambda_2)$ and (μ_0, μ_1, μ_2) such that:

$$R(x) = \lambda_0\phi_0(x) + \mu_0\phi_1(x) + \lambda_1\phi_2(x) + \mu_1\phi_3(x) + \lambda_2\phi_4(x) + \mu_2\phi_5(x)$$

has class C^2 and correctly interpolates f.

Necessarily, $(\lambda_0, \lambda_1, \lambda_2) = (f(1), f(3), f(5))$.

We need to ensure that the function is continuous at the point 3:

$$\mu_0 + 4\mu_1 + \mu_2 = \frac{3}{h}(f(5) - f(1)).$$

The second derivatives vanish at the endpoints of the interval: $2\mu_0 + \mu_1 = \dfrac{3}{h}(f(3) - f(1))$ and $\mu_1 + 2\mu_2 = \dfrac{3}{h}(f(5) - f(3))$.

Finally:

$$R(x) = \frac{1}{2}\phi_0(x) - \frac{63}{260}\phi_1(x) + \frac{1}{10}\phi_2(x) - \frac{3}{26}\phi_3(x) + \frac{1}{26}\phi_4(x) - \frac{3}{260}\phi_5(x).$$

We need to find ϕ_i, $i = 1, \ldots, 5$.

We can now find $R(4)$ in both cases and compare with $f(4)$.

Thus, $R(4) = \dfrac{3}{80} = 0.0375$. Here, the error is:

$$f(4) - R(4) = \frac{1}{17} - \frac{3}{80} \simeq 2.13 \times 10^{-2}.$$

3.7. Using *Matlab*

3.7.1. *Operations on polynomials*

In *Matlab*, polynomials are represented by row vectors whose components are ordered by decreasing powers. A polynomial of degree n is represented by a vector of size $(n+1)$. For example, the polynomial $f(x) = 2x^3 - 3x^2 + x - 1$ is represented by:

```
>>f=[2 -3 1 -1]
f =
     2 -3 1 -1
```

3.7.1.1. *Polynomial multiplication*

The function *conv* returns the convolution product of two polynomials. The following example demonstrates how to use this function.

Consider the two polynomials $f(x) = 3x^3 + 2x^2 - x + 4$ and $g(x) = 2x^4 - 3x^2 + 5x - 1$. Their convolution product $h(x) = f(x) * g(x)$ is returned by:

```
>>f=[3 2 -1 4];

>>g=[2 0 -3 5 -1];

>>h=conv(f,g)
h =
     6 4 -11 17 10 -19 21 -4
```

This returns the polynomial $h(x) = 6x^7 + 4x^6 - 11x^5 + 17x^4 + 10x^3 - 19x^2 + 21x - 4$.

3.7.1.2. *Polynomial division*

The function *deconv* returns the convolution quotient of two polynomials (deconvolution of the coefficients of the polynomial). The following example demonstrates how to use this function.

Consider the same polynomials as before, $f(x)$ and $g(x)$. The quotient of $g(x)$ by $f(x)$ is returned by the function *deconv*:

```
>>f=[3 2 -1 4];

>>g=[2 0 -3 5 -1];

>>h=deconv(g,f)
h =
     0.6667 -0.4444
```

This returns the polynomial $h(x) = 0.6667x - 0.4444$.

3.7.2. Manipulating polynomials

In *Matlab*, the polynomial $P(x) = 2x^3 + x^2 + 4x + 5$ is represented by:

```
>>P=[2 1 4 5];
```

This polynomial has roots r_i. To find these roots, we can call the *roots* function. Thus:

```
>>r=roots(P);

>> r
r =
      0.2500 + 1.5612i
      0.2500 - 1.5612i
     -1.0000
```

The three roots of this polynomial (two of which are complex) are returned as a column vector. For a given set of roots r_i, the coefficients of the polynomial with these roots can be calculated using the *poly* command:

```
>>poly(r)
ans =
     1.0000 0.5000 2.0000 2.5000
```

The *poly* function also accepts matrix inputs, in which case it returns the characteristic polynomial of the input. For example:

```
>>A=[3 1;2 4];

>>p=poly(A)
p =
     1 -7 10
```

Thus, the characteristic polynomial of the matrix A is $p(x) = x^2 - 7x + 10$.

The roots of this polynomial are the eigenvalues of the matrix A. These roots are returned by the function *eig*:

```
>>eigval=eig(A)
eigval =
         2
         5
```

3.7.3. *Evaluation of polynomials*

To evaluate a polynomial at a given point, we can use the *polyval* function. For example, to evaluate the polynomial p at $x = 1$:

```
>>polyval(P,1)
ans =
    12
```

3.7.4. *Linear and nonlinear interpolation*

Interpolation uses a piecewise linear or polynomial curve to connect experimentally obtained points together. This can be done automatically with the "interp1" function. The command "interp1(x,y,xi,'type')" returns a vector with the same dimensions as x_i whose components are the images of the components of x_i, calculated by interpolating from the points x and y. We write f for the interpolant of y. The "type" string specifies the type of interpolation, which can take the following values:

– "linear": linear interpolation;

– "spline": cubic spline interpolation;

– "cubic": cubic interpolation.

If the type is left unspecified, linear interpolation is performed by default.

EXAMPLE.– Suppose that we wish to interpolate the function $\sin x$ on $n = 2, 3, 4, 5, 6$ nodes. In *Matlab*, we can use the command *polyfit* to calculate the coefficients of the interpolating polynomial and *polyval* to evaluate a polynomial whose coefficients are known at a sequence of points. Executing the program listed below yields the following diagram:

```
f='sin(x)'; x=[0:3*pi/100:3*pi];
x_sample=x;
plot(x,eval(f),'b');
hold on
for i=2:6
```

```
    x = linspace(0,3*pi,i);
    y= eval(f);
    c=polyfit(x,y,i-1);
    plot(x_sample,polyval(c,x_sample),'b--')
end
```

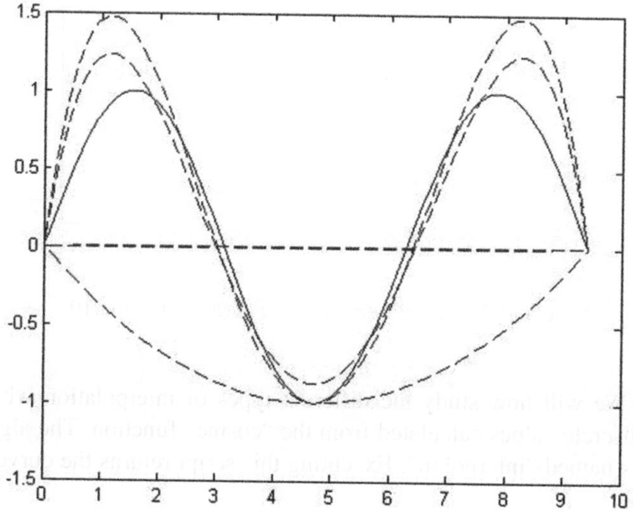

Figure 3.6. *Interpolation of* $\sin x$ *between* 0 *and* 3π

EXAMPLE.– Executing the commands listed below returns Figure 3.7.

```
>>x = 0:10;
>>y = sin(x);
>>xi = 0:.25:10;
>>yi = interp1(x,y,xi,'cubic');
>>plot(x,y,'o',xi,yi)
```

Matlab has a built-in function called “spline” for spline interpolation. For example, returning to the example of the program for the sine function above, we can replace the yi with the following command in order to implement spline interpolation:

```
>> yi = spline(x,y,xi);
```

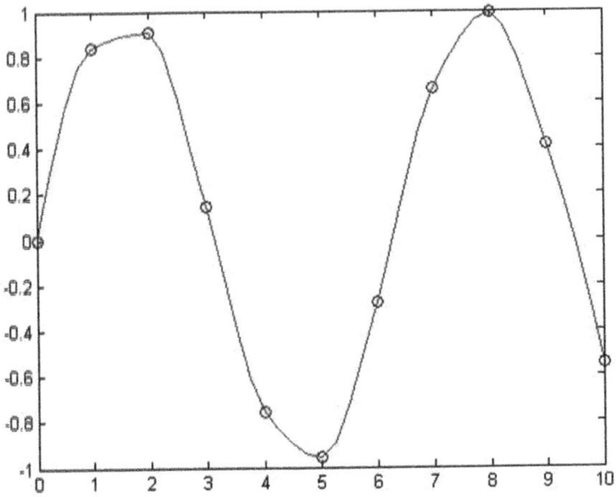

Figure 3.7. *Interpolation of* $\sin x$ *between* 0 *and* 10

EXAMPLE.– We will now study the different types of interpolation with a similar example of discrete values calculated from the "cosine" function. The algorithm that we will run is named "interpol.m". Executing this script returns the curves shown in Figure 3.8.

```
clear all; clc;
x=0:10;
y=cos(x);     % Points to be interpolated
z=0:0.25:10; % z has a smaller step size than x

% Linear interpolation
figure(1);
f=interp1(x,y,z);

% Plot of the true values and the interpolated curve
plot(x,y,'*r',z,f);
grid on; xlabel('Interpolation');

% Interpolation by cubic splines
figure(2);
f=interp1(x,y,z,'spline');
plot(x,y,'*r',z,f);
grid on;
xlabel('Interpolation by cubic splines');
```

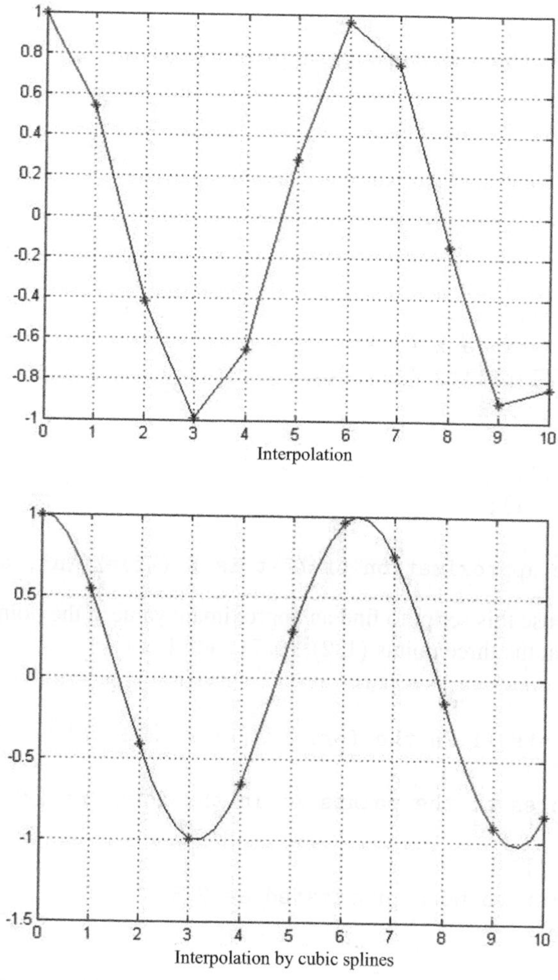

Figure 3.8. *Interpolation of* $\cos x$ *between* 0 *and* 10

The function *interp2* performs interpolation in three-dimensional (3D) space.

3.7.5. *Lagrange function*

The following *Matlab* script can be used to perform Lagrange interpolation:

```
x=input('Enter the points Xi in the form Xi=[x1 x2 ... xn]: X(i)=');

y=input('Enter the images of the points Xi in the form Yi= [y1y2...
```

```
yn]: Y (i)= ');

a=input('Enter the point to be approximated a= ');

n=length(x);
p=0;
for i=1:n
    pr=1;
    prl=1;
    for j=1:n
        if(j~=i)
            pr=pr*(a-x (j));
            prl=prl*(x (i)-x (j));
        end
    end
    L(i)=pr/prl;
    p=p+y(i)*L(i);
end
sprintf ('The approximation of %f\t is f (%f)=%f\n', a, a, p)
```

We will now use this script to find an approximate value at the point $a = 2.5$ for a function known at the three points $(1, 2)$, $(3, 7)$ and $(4, -1)$.

```
>> Lagrange
Enter the points Xi in the form Xi=[x1 x2 ... xn]:
X (i)= [1 3 4]
Enter the images of the points Xi in the form Yi= [y1 y2 ... yn]:

Y (i)= [2 7 -1]
Enter the point to be approximated a= 2.5

ans =
    The approximation of 2.500000 is f (2.500000)=8.375000
```

3.7.6. *Newton function*

The following script calculates the Newton polynomial:

```
clc; clear;
n=input('specify the number of points to interpolate,n= ');
for i=1:n
       A(1,i)=input('value of Xi= ');
       A(2,i)=input('value of f(Xi)= ');
end
i=1;
```

```
for k=3:n+1
    for j=1:n-i
        A(k,j)=[A(k-1,j+1)-A(k-1,j)]/[A(1,j+i)-A(1,j)];
    end
    i=i+1;
end
x = sym('x');
P=A(2,1);
L=1;
for i=1:n-1
    L=L*(x-A(1,i));
    P=P+L*A(i+2,1);
end
disp('P='), disp(P);
```

We can apply this script to calculate the Newton polynomial for a function that is known at the points $(1, 2)$, $(3, 7)$ and $(4, -1)$ as follows:

```
specify the number of points to interpolate,n= 3
value of Xi= 1

value of f(Xi)= 2
value of Xi= 3
value of f(Xi)= 7
value of Xi= 4
value of f(Xi)=-1
P= -1/2+5/2*x-7/2*(x-1)*(x-3)
```

4

Numerical Differentiation

4.1. First-order numerical derivatives and the truncation error

Let $f \in \mathcal{C}^1(\mathbb{R}, \mathbb{R})$. If $x_0 \in \mathbb{R}$, then:

$$f'(x_0) = \lim_{h \to 0} \frac{f(x_0 + h) - f(x_0)}{h} \quad [4.1]$$

$$= \lim_{h \to 0} \frac{f(x_0) - f(x_0 - h)}{h} \quad [4.2]$$

$$= \lim_{h \to 0} \frac{f(x_0 + \frac{h}{2}) - f(x_0 - \frac{h}{2})}{h}. \quad [4.3]$$

One approach to numerically computing $f'(x_0)$ is therefore to choose a sufficiently small $h > 0$ and calculate:

$$\frac{\triangle_h f(x_0)}{h} \text{ or } \frac{\nabla_h f(x_0)}{h} \text{ or } \frac{\delta_h f(x_0)}{h}$$

after defining the quantities:

$$\triangle_h f(x_0) = f(x_0 + h) - f(x_0) \quad [4.4]$$

$$\nabla_h f(x_0) = f(x_0) - f(x_0 - h) \quad [4.5]$$

$$\delta_0 f(x_0) = f\left(x_0 + \frac{h}{2}\right) - f\left(x_0 - \frac{h}{2}\right). \quad [4.6]$$

For any given $h > 0$:

$$\begin{aligned} \triangle_h : \mathcal{C}(\mathbb{R}, \mathbb{R}) &\to \mathcal{C}(\mathbb{R}, \mathbb{R}) \\ f &\mapsto \triangle_h f(x) = f(x + h) - f(x). \end{aligned}$$

Similar statements hold for the operators ∇_h and δ_h.

DEFINITION.– *For any given $h > 0$, the operators Δ_h, ∇_h and δ_h are known as the first-order forward, backward and centered difference operators respectively.*

These operators satisfy the following result:

THEOREM 4.1.– *The first-order difference operators Δ_h, ∇_h and δ_h are linear.*

For $f \in C^2(\mathbb{R}, \mathbb{R})$, the second-order Taylor expansion of f about the point x_0 may be written as

$$f(x_0 + h) = f(x_0) + f'(x_0)h + \frac{1}{2}f''(\xi)h^2, \qquad [4.7]$$

where $\xi \in [x_0, x_0 + h]$. From [4.4] and [4.7], we deduce

$$\left| f'(x_0) - \frac{\Delta_h f(x_0)}{h} \right| = \frac{1}{2}|f''(\xi)|h. \qquad [4.8]$$

EXAMPLE.– Let f be a regular function whose value is known at three equally spaced points $x_{-1} = x_0 - h$, x_0, and $x_1 = x_0 + h$. Then:

$$f'x_0 = \frac{1}{2h}[f(x_1) - f(x_{-1})] - \frac{h^2}{6}f^{(3)}(\xi), \quad \xi \in [x_{-1}, x_1].$$

To prove this, we first find the Lagrange interpolating polynomial of f for the three equally spaced points x_{-1}, x_0, and x_1:

$$P_2(x) = L_{-1}(x)f(x_{-1}) + L_0(x)f(x_0) + L_1(x)f(x_1)$$

$$L_{-1}(x) = \frac{(x - x_0)(x - x_1)}{(x_{-1} - x_0)(x_{-1} - x_1)} = \frac{(x - x_0)(x - x_1)}{2h^2}$$

$$L_0(x) = \frac{(x - x_{-1})(x - x_1)}{(x_0 - x_{-1})(x_0 - x_1)} = -\frac{(x - x_1)(x - x_1)}{h^2}$$

$$L_1(x) = \frac{(x - x_{-1})(x - x_0)}{(x_1 - x_1)(x_1 - x_0)} = \frac{(x - x_{-1})(x - x_0)}{2h^2}.$$

Differentiating the Lagrange polynomials yields

$$L'_{-1}(x) = \frac{1}{2h^2}[(x - x_0) + (x - x_1)] \Rightarrow L'_{-1}(x_0) = -\frac{1}{2h}$$

$$L'_0(x) = -\frac{1}{h^2}[(x - x_1) + (x - x_1)] \Rightarrow L'_0(x_0) = 0$$

$$L'_1(x) = \frac{1}{2h^2}[(x - x_{-1}) + (x - x_0)] \Rightarrow L'_1(x_0) = \frac{1}{2h}$$

Therefore,

$$P'_2(x_0) = \frac{1}{2h}[f(x_1) - f(x_{-1})].$$

The derivative of the residual is

$$E'(x_0) = \frac{1}{6}(x_0 - x_{-1})(x_0 - x_1)f^{(3)}(\xi) = -\frac{h^2}{6}f^{(3)}(\xi),$$

which implies that

$$f'(x_0) = \frac{1}{2h}[f(x_1) - f(x_{-1})] - \frac{h^2}{6}f^{(3)}(\xi).$$

We therefore have the following results:

THEOREM 4.2.– *Given $f \in C^2(\mathbb{R}, \mathbb{R})$, $x_0 \in \mathbb{R}$, and $h_0 > 0$, there exists a constant C such that*

$$\left| f'(x_0) - \frac{\Delta_h f(x_0)}{h} \right| \leq Ch, \quad \forall h \leq h_0. \qquad [4.9]$$

THEOREM 4.3.– *Given $f \in C^3(\mathbb{R}, \mathbb{R})$, $x_0 \in \mathbb{R}$, and $h_0 \in \mathbb{R}_+^*$, there exists a constant C such that*

$$\left| f'(x_0) - \frac{\nabla_h f(x_0)}{h} \right| \leq Ch^2, \quad \forall h \leq h_0. \qquad [4.10]$$

Theorems 4.2 and 4.3 guarantee that the quantities $\frac{\Delta_h f(x_0)}{h}$ and $\frac{\nabla_h f(x_0)}{h}$ will converge to $f'(x_0)$ as h tends to 0 whenever f is sufficiently regular. In the first case, the convergence is of order h; in the second case, the convergence is of order h^2.

DEFINITION.– *We say that $\frac{\Delta_h f(x0)}{h}$ and $\frac{\nabla_h f(x_0)}{h}$ are the forward and backward finite difference formulas that approximate $f'(x_0)$, and the differences:*

$$\left| f'(x_0) - \frac{\Delta_h f(x_0)}{h} \right| \text{ and } \left| f'(x_0) - \frac{\nabla_h f(x_0)}{h} \right| \qquad [4.11]$$

are the corresponding truncation errors, which are of order h. We say that the finite difference formulas are consistent up to the first order in h.

Similarly, the centered finite differences formula $\frac{\delta_h f(x_0)}{h}$ that approximates $f'(x_0)$ is consistent up to the second order in h, since its truncation error is of order h^2. Therefore, the centered finite differences formula is more accurate than the forward and backward versions.

4.2. Higher-order numerical derivatives

Let $n \in \mathbb{N}^*$. We recursively define:

$$\Delta_h^n f = \Delta_h(\Delta_h^{n-1} f), \tag{4.12}$$

$$\nabla_h^n f = \nabla_h(\nabla_h^{n-1} f), \tag{4.13}$$

$$\delta_h^n f = \delta_h(\delta_h^{n-1} f). \tag{4.14}$$

Thus, for example:

$$\begin{aligned}
\delta_h^2 f(x) &= \delta_h(\delta_h f(x)) = \delta_h\left(f\left(x+\frac{h}{2}\right) - f\left(x-\frac{h}{2}\right)\right) \\
&= \delta_h f\left(x+\frac{h}{2}\right) - \delta_h f\left(x-\frac{h}{2}\right) \\
&= f(x+h) - f(x) - \left[f\left(x-\frac{h}{2}+\frac{h}{2}\right) - f\left(x-\frac{h}{2}-\frac{h}{2}\right)\right] \\
&= f(x+h) - 2f(x) + f(x-h),
\end{aligned}$$

similar to the case where $n = 1$. It can be verified that the operators Δ_h^n, ∇_h^n, and δ_h^n are linear.

It is easy to prove that if f is a sufficiently regular function ($f \in \mathcal{C}^{n+1}$ for forward or backward finite differences or $f \in \mathcal{C}^{n+2}$ for centered finite differences) and $x_0 \in \mathbb{R}$, then the values:

$$\frac{\Delta_h^n f(x_0)}{h^n}, \quad \frac{\nabla_h^n f(x_0)}{h^n}, \quad \frac{\delta_h^n f(x_0)}{h^n}$$

are approximations of $f^{(n)}(x_0) = \dfrac{d^n}{dx^n} f(x_0)$ of order h, h, and h^2 respectively as $h \to 0$.

This allows us to state the following results, which generalize Theorems 4.2 and 4.3.

THEOREM 4.4.– *For given $n \in \mathbb{N}^*$, $f \in \mathcal{C}^{n+1}(\mathbb{R}, \mathbb{R})$, $x_0 \in \mathbb{R}$, and $h_0 > 0$, there exists a constant C such that:*

$$\left| f^{(n)}(x_0) - \frac{\Delta_h^n f(x_0)}{h^n} \right| \leq Ch \quad \forall h \leq h_0 \tag{4.15}$$

$$\left| f^{(n)}(x_0) - \frac{\nabla_h^n f(x_0)}{h^n} \right| \leq Ch \quad \forall h \leq h_0. \tag{4.16}$$

THEOREM 4.5.– *For any given $n \in \mathbb{N}^*$, $f \in \mathcal{C}^{n+2}(\mathbb{R}, \mathbb{R})$, $x_0 \in \mathbb{R}$, and $h_0 > 0$, there exists a constant C such that:*

$$\left| f^{(n)}(x_0) - \frac{\delta_h^n f(x_0)}{h^n} \right| \leq Ch^2 \quad \forall h \leq h_0. \qquad [4.17]$$

Diffusion problems, elastic deformation problems, wave propagation problems, fluid flow problems and so on, all have terms with second- and fourth-order derivatives. Therefore, the centered finite difference formulas that approximate $f''(x_0)$ $(n = 2)$ and $f^{(4)}(x_0)$ $(n = 4)$ are very frequently used by engineers. These formulas may be stated as follows:

$$f''(x_0) \approx \frac{f(x_0 + h) - 2f(x_0) + f(x_0 + h)}{h^2} \qquad [4.18]$$

$$f^{(4)}(x_0) \approx \frac{f(x_0 + 2h) - 4f(x_0 + h) + 6f(x_0) - 4f(x_0 - 2h) + f(x_0 - 2h)}{h^4} \qquad [4.19]$$

REMARK.– It is important to note that numerical differentiation is a highly unstable operation. This means that it is extremely sensitive to rounding errors as n increases.

4.3. Numerical derivatives and interpolation

Let $f \in \mathcal{C}(\mathbb{R}, \mathbb{R})$, $x_0 \in \mathbb{R}$, and suppose that $h > 0$ is sufficiently small. Consider the points $x_j = x_0 + jh$, for $j = 0, 1, 2, \ldots$. Given $n \in \mathbb{N}^*$, we can construct the Newton polynomial $P_n(x)$ as follows:

$$\begin{aligned} P_n(x) &= f(x_0) + \frac{\Delta_h f(x_0)}{h}(x - x_0) + \cdots \\ &\quad + \frac{\Delta_h^n f(x_0)}{n!h^n}(x - x_0)(x - x_1)\ldots(x - x_{n-1}). \end{aligned} \qquad [4.20]$$

P_n is a polynomial of degree n, and $P_n(x_0) = f(x_0)$:

$$\begin{aligned} P_n(x_1) &= f(x_0) + \frac{\Delta_h f(x_0)}{h}(x_1 - x_0) = f(x_0) + \frac{f(x_1) - f(x_0)}{h}.h = f(x_1), \\ P_n(x_2) &= f(x_0) + \frac{\Delta_h f(x_0)}{h}(x_2 - x_0) + \frac{\Delta_h^2 f(x_0)}{2h^2}(x_2 - x_0)(x_2 - x_1) \\ &= f(x_0) + 2\Delta_h f(x_0) + \Delta_h^2 f(x_0) \\ &= f(x_0) + 2(f(x_1) - f(x_0)) + (f(x_2) - 2f(x_1)) + f(x_0) = f(x_2). \end{aligned}$$

In fact, it can be shown that $P_n(x_j) = f(x_j)$, $j = 0, 1, 2, \ldots, n$. Furthermore, since P_n is a polynomial of degree n, it is unique such that the polynomial interpolates f on the $(n + 1)$ points $x_0, x_1, x_2, \ldots, x_n$. It is easy to note that:

$$\frac{d^n}{dx^n} P_n(x) = \frac{\Delta_h^n f(x_0)}{h^n}.$$

This partially proves the following result:

THEOREM 4.6.– *Let P_n be the polynomial of degree n that interpolates f on the points $x_j = x_0 + jh$ for $j = 0, 1, 2, \ldots, n$. Then:*

$$P_n(x) = f(x_0) + \frac{\Delta_h f(x_0)}{h}(x - x_0) + \cdots + \frac{\Delta_h^n f(x_0)}{n!h^n}(x - x_0)(x - x_1)\ldots(x - x_{n-1}) \quad [4.21]$$

and:

$$\frac{d^n}{dx^n}P_n(x) = \frac{\Delta_h^n f(x_0)}{h^n}.$$

REMARK.– If $f \in \mathcal{C}^{n+1}(\mathbb{R}, \mathbb{R})$, we can use Theorem 4.6 to find an expression for the following error between f and P_n:

$$\max_{x_0 \le x \le x_0+nh} |f(x) - P_n(x)| \le \frac{1}{2(n+1)}h^{n+1} \max_{x_0 \le x \le x_0+nh} |f^{(n+1)}(x)|.$$

REMARK.– Results similar to those stated in Theorem 4.6 also hold for the operators ∇_h^n and δ_h^n. For example, it can easily be shown that if P_2 is the quadratic polynomial that interpolates the function f at the points $x_0 - h$, x_0, and $x_0 + h$, then:

$$\frac{d^2}{dx^2}P_2(x_0) = \frac{\delta_h^2 f(x_0)}{h^2} = \frac{f(x_0 + h) - 2f(x_0) + f(x_0 - h)}{h^2}.$$

Moreover, we know that $f(x) = P_n(x) + E(x)$. *A priori*, $f^{(p)}(x) = P_n^{(p)}(x) + E^{(p)}(x)$, where:

$$P_n(x) = \sum_{i=0}^{n} \phi_i(x) f(x_i) \Rightarrow f^{(p)}(x) = \sum_{i=0}^{n} \phi_i^{(p)}(x) f(x_i) + E^{(p)}(x).$$

In order for $P_n^{(p)}(x)$ to be a good approximation of $f^{(p)}(x)$, we need $E(x)$ to be p-times differentiable, and the p-th derivative must be negligibly small relative to $P_n^{(p)}(x)$. This means that we need to study $E(x)$.

EXAMPLE.– Let f be a regular function whose values are known at three points equally spaced around x_0. Then:

$$f'(x_0) = \frac{1}{2h}[f(x_1) - f(x_{-1})] - \frac{h^2}{6}f^{(3)}(\xi),$$

where $\xi \in [x_{-1}, x_1]$. This is known as the central difference formula.

Let P_2 be the Lagrange interpolating polynomial on these three points. Then:

$$P_2(x) = L_{-1}(x)f(x_{-1}) + L_0(x)f(x_0) + L_1(x)f(x_1),$$

where:

$$L_{-1}(x) = \frac{(x-x_0)(x-x_1)}{(x_{-1}-x_0)(x_{-1}-x_1)} = \frac{(x-x_0)(x-x_1)}{2h^2}$$

$$L_0(x) = \frac{(x-x_{-1})(x-x_1)}{(x_0-x_{-1})(x_0-x_1)} = \frac{(x-x_{-1})((x-x_1))}{h^2}$$

$$L_1(x) = \frac{(x-x_{-1})(x-x_0)}{2h^2},$$

with derivatives:

$$L'_{-1}(x) = \frac{1}{2h^2}[(x-x_0)+(x-x_1)] \Rightarrow L'_{-1}(x_0) = -\frac{1}{2h}$$

$$L'_0(x) = \frac{1}{h^2}[(x-x_{-1})+(x-x_1)] \Rightarrow L'_0(x_0) = 0$$

$$L'_1(x) = \frac{1}{2h}[(x-x_{-1})+(x-x_0)] \Rightarrow L'_1(x_0) = \frac{1}{2h}.$$

Therefore:

$$P'_2(x_0) = \frac{1}{2h}[f(x_1)-f(x_{-1})],$$

and the derivative of the interpolation error may be written as:

$$\begin{aligned} E'(x_0) &= \prod_{j\neq 0}(x_0-x_j).\frac{1}{3!}f^{(3)}(\xi) \quad \xi \in [x_{_1}, x_1] \\ &= \frac{1}{6}(x_0-x_{-1})(x_0-x_1)f^{(3)}(\xi) \\ &= -\frac{h^2}{6}f^{(3)}(\xi). \end{aligned}$$

Since $f'(x_0) = P'_2(x_0) + E'(x_0)$, we have that $f'(x_0) = \frac{1}{2h}[f(x_1) - f(x_{-1})] - \frac{h^2}{6}f^{(3)}(\xi)$.

4.4. Studying the differentiation error

We shall begin by studying $E'(x) = f'(x) - P'(x)$. If f is $(n+1)$-times continuously differentiable on $]a, b[$, then, for all x in $[a, b]$, there exists ξ_x in $[a, b]$ such that:

$$E(x) = \frac{1}{(n+1)!}\pi(x)f^{(n+1)}(\xi_x), \text{ where } \pi(x) = \prod_{i=0}^{n}(x-x_i).$$

We can therefore write $E(x) = \pi(x)g(x)$, where $g(x) = \frac{1}{(n+1)!}f^{(n+1)}(\xi_x)$. By formally differentiating this expression, $E'(x) = \pi'(x)g(x) + \pi(x)g'(x)$.

The point at which we wish to approximate f' might be:

– one of the points x_i, $i = 0, \ldots, n$;

– a point distinct from all of the x_i, $\forall i = 0, \ldots, n$.

a) If $x = x_i$, then $E'(x_i) = \pi'(x_i)g(x_i)$, where:

$$\pi'(x_i) = \lim_{x \to x_i} \frac{\pi(x) - \pi(x_i)}{x - x_i} = \lim_{x \to x_i} \frac{\pi(x)}{x - x_i} = \prod_{\substack{j=0 \\ j \neq i}}^{n} (x_i - x_j),$$

and:

$$E'(x_i) = \frac{1}{(n+1)!} \prod_{\substack{j=0 \\ j \neq i}}^{n} (x_i - x_j) f^{(n+1)}(\xi_{x_i}).$$

b) If $x \neq x_i$, $\forall i = 1, \ldots, n+1$, we need an estimate of $g'(x)$, so we need to assume that f is $(n+2)$-times differentiable. For all x in $[a, b]$, there exists ξ_x such that $g'(x) = \frac{1}{(n+2)!} f^{(n+2)}(\xi_x)$, so:

$$E'(x) = \frac{1}{(n+1)!}\pi(x) f^{(n+1)}(\delta_x) + \frac{1}{(n+2)!}\pi(x) f^{(n+2)(\xi_x)}.$$

More generally:

$$E(x) = \pi(x)g(x) \Rightarrow E^{(p)}(x) = \sum_{m=0}^{p} C_p^m \pi^{(p-m)}(x) g^{(m)}(x).$$

From the Newton interpolating polynomial and divided differences formula:

$$\begin{aligned} E(x) &= f(x) - P(x) \\ &= (x - x_0)(x - x_1) \ldots (x - x_n) y(x, x_0, x_1, \ldots, x_n) \\ &= \pi(x) y(x, x_0, x_1, \ldots, x_n). \end{aligned}$$

If $y(x)$ has class C^1, then:

$$E'(x) = \pi'(x) y(x, x_0, x_1, \ldots, x_n) + \pi(x) \frac{d}{dx} y(x, x_0, x_1, \ldots, x_n).$$

However:

$$\frac{d}{dx}y(x, x_0, x_1, \ldots, x_n) = \lim_{h\to 0} \frac{y(x+h, x_0, x_1, \ldots, x_n) - y(x, x_0, x_1, \ldots, x_n)}{h}$$
$$= \lim_{h\to 0} y(x+h, x, x_0, x_1, \ldots, x_n)$$
$$= y(x, x, x_0, x_1, \ldots, x_n),$$

which implies that:

$$E'(x) = \pi'(x)y(x, x_0, x_1, \ldots, x_n) + \pi(x)y(x, x, x_0, x_1, \ldots, x_n).$$

Moreover:

$$y(x, x_0, x_1, \ldots, x_n) = \frac{y^{(n+1)}(c_x)}{(n+1)!} \qquad c_x \in [a, b].$$

Therefore:

$$E'(x) = \pi'(x)\frac{y^{(n+1)}(c_x)}{(n+1)!} + \pi(x)\frac{y^{(n+2)}(\alpha_x)}{(n+2)!} \quad c_x, \alpha_x \in [a, b],$$

provided that $y(x)$ is $(n+2)$-times continuously differentiable.

EXAMPLE.– Consider the function $f\colon [1, 3] \to \mathbb{R}$ defined by $f(x) = \frac{1}{x}$, and the points $x_0 = 1$, $x_1 = 2$ and $x_3 = 3$.

We can calculate approximate values of the real numbers $f'(2)$ and $f'(\frac{3}{2})$ using the Lagrange polynomial of the sequence (x_0, x_1, x_2).

By the example given above, we can directly write that:

$$f'(2) = \frac{1}{2h}[f(3) - f(1)] - \frac{h^2}{6}f^{(3)}(\xi); \quad h = 1$$
$$= \frac{1}{2}\left[\frac{1}{3} - 1\right] - \frac{1}{6}\left(-\frac{6}{\xi^4}\right) = -\frac{1}{3} + \frac{1}{\xi^4}.$$

The interpolating polynomial of f at the points 1, 2, and 3 is $P_2(x) = L_1(x)f(1) + L_2(x)f(2) + L_3(x)f(3)$, where:

$$L_1(x) = \frac{(x-2)(x-3)}{(1-2)(1-3)} = \frac{x^2 - 5x + 6}{2},$$

$$L_2(x) = \frac{(x-1)(x-3)}{(2-1)(2-3)} = -x^2 + 4x - 3,$$

$$L_3(x) = \frac{(x-1)(x-2)}{(3-1)(3-2)} = \frac{x^2 - 3x + 2}{2}.$$

Therefore:

$$P_2(x) = \frac{x^2 - 6x + 11}{6},$$

which implies that $P_2'(x) = \dfrac{x}{3} - 1$. Hence, $f'(\frac{3}{2}) \approx \frac{3}{2}\frac{1}{3} - 1 = -\frac{1}{2}$.

To find an upper bound for the error:

– for $x = 2$,

$$|E'(2)| \leq \frac{1}{6} \sup_{\xi \in [1,3]} |f^{(3)}(\xi)| \leq \sup_{x \in [1,3]} \left| \frac{1}{x^4} \right| = 1.$$

This is a very rough estimate.

– for $x = \frac{3}{2}$,

$$E'\left(\frac{3}{2}\right) = \frac{1}{3!}\pi'\left(\frac{3}{2}\right) f^{(3)}(\xi_{3/2}) + \frac{1}{4!}\pi\left(\frac{3}{2}\right) f^{(4)}(\delta_{3/2}).$$

Since $\pi(x) = x^3 - 6x + 12x - 6$, we have that $\pi'(x) = 3x^2 - 12x + 11$, $f^{(3)}(x) = \frac{6}{14}$ and $f^{(4)} = \frac{24}{x^5}$. Therefore, $\pi\left(\dfrac{3}{2}\right) = \dfrac{3}{8}$ and $\pi'\left(\dfrac{3}{2}\right) = -\dfrac{1}{4}$, so:

$$\begin{aligned} |E'(\frac{3}{2})| &\leq \sup_{x \in [1,3]} |\frac{1}{x^4}| \frac{3}{8} + |-\frac{1}{4}| \sup_{x \in [1,3]} |\frac{1}{x^5}| \\ &\leq \frac{3}{8} + \frac{1}{4} = \frac{5}{8}. \end{aligned}$$

Another approach to estimating the error is to write $E(x) = \pi(x)y(x, x_0, x_1, x_2) = \pi(x)y(x, 1, 2, 3)$, which implies that:

$$E'(x) = \pi'(x)y(x, 1, 2, 3) + \pi(x)y(x, x, 1, 2, 3).$$

We can compute $y(x, 1, 2, 3)$ using divided differences, which allows us to find $y(x, 1, 2, 3) = -\frac{1}{6x}$. For $x = x_i$, we can take the limit $\frac{dy(x,1,2,3)}{dx} = \frac{1}{6x^2}$.

For $x = 2$, $E^2 = \pi'(2)y(2, 1, 2, 3) = -1 \times (-\dfrac{1}{12}) = 8.3 \times 10^{-2}$.

For $x = \dfrac{3}{2}$, $E'\left(\dfrac{3}{2}\right) = \pi'\left(\dfrac{3}{2}\right) y(\dfrac{3}{2}, 1, 2, 3) + \pi\left(\dfrac{3}{2}\right) y(x, x, 1, 2, 3) = 5.5 \times 10^{-2}$.

4.5. Richardson extrapolation

The accuracy of the computation of the derivative is limited by two cumulative phenomena: errors at the level of the numerical scheme itself, and machine representation errors. As a result, the accuracy does not tend to infinity as h tends to zero, which some readers may find counterintuitive. In other words, there exists an optimal choice h_{opt} of h that achieves an accuracy threshold that cannot be exceeded.

One powerful method for increasing the order of a numerical differentiation scheme is known as Richardson extrapolation. This method improves the accuracy by calculating the derivative with multiple step sizes:

$$h_0 \equiv h, \; h_1 \equiv \frac{h}{2}, \; h_2 \equiv \frac{h_0}{2^2}, \; \dots, \; h_n \equiv \frac{h_0}{2^n}$$

whenever possible, and combining these values. For instance, by writing $f'[h_k]$ for the derivative at a point computed using the step size $h_k = \frac{h_0}{2^k}$, the formula:

$$f' \approx \frac{2^k f'[h_{k+1}] - f'[h_k]}{2^k - 1}$$

gives an improved approximation. In practice, this increases the order of the numerical method by at least one.

EXAMPLE.– Suppose that we wish to compute an estimate of $u''(0)$ at the origin using only the values listed in the following table (rounded to three decimal places):

x	$-2h$	$-h$	0	h	$2h$
$f(x)$	0.418	0.423	0.426	0.424	0.423

We have that:

$$\Delta_{2h} f = \frac{f(-2h) - 2f(0) + f(2h)}{4h^2} = \frac{-0.0025}{h^2},$$
$$\Delta_h f = \frac{f(-h) - 2f(0) + f(h)}{h^2} = \frac{-0.003}{h^2},$$

which implies that:

$$\Delta_{\text{Richardson}} f = \frac{4\Delta_h f - \Delta_{2h} f}{3} = \frac{-0.0120 + 0.0025}{3h^2} \approx \frac{-0.0032}{h^2}.$$

The linear combination eliminates the first term of the discretization error.

First, we find an expression for the discretization error by considering the corresponding linear combinations of the errors for Δ_{2h} and Δ_h. Thus:

$$\frac{1}{3}\left[4\left(\frac{C_1 h^2}{12} + \frac{C_2 h^4}{360}\right) - \left(\frac{C_1 h^2}{12} + \frac{C_2 h^4}{360}\right)\right] = -\frac{C_2 h^4}{90}.$$

Next, we derive an expression for the rounding error by propagating the rounding errors of Δ_{2h} and Δ_h in the Richardson linear combination. These errors are cumulative, so:

$$\frac{1}{3}\left[4\frac{4\epsilon}{h^2}+\frac{\epsilon}{h^2}\right]=\frac{17\epsilon}{3h^2}.$$

We deduce the following bound for the computation error:

$$|E_{\text{Richardson}}| \leq \frac{17\epsilon}{3h^2}+\frac{C_2h^4}{90}.$$

We can now simply minimize $g(h)=\frac{17\epsilon}{3h^2}+\frac{h^4}{90}$ by requiring:

$$g'(h)=\frac{-34\epsilon}{3h^3}+\frac{4h^3}{90}=0,$$

which implies that:

$$h^6=\frac{45}{2}\frac{34}{3}\epsilon=255\epsilon.$$

This leads us to select the step size $h=\sqrt[6]{255\epsilon}$.

4.6. Application to the heat equation

We will now study the dissipation of heat from a radiating element (or fin). We will assume that the fin has length L and is sufficiently thin that the problem is one-dimensional (i.e. the temperature T only depends on x). Thus, T satisfies the following differential equation:

$$-kT''(x)+\frac{hc(x)p}{S}(T-T_e)=0 \quad \forall x\in\Omega=]0,L[,$$

where k is the thermal conductivity, $h_c(x)$ is the surface heat transfer coefficient, p is the perimeter of the surface S (pdx is the lateral surface) and T_e is the ambient temperature.

By writing $u(x)=T(x)-T_e$ and assuming that the fin is heated to temperature T_0 at both ends to provide boundary conditions, the problem may be expressed as:

$$\begin{cases} -u''(x)+c(x)u(x)=0 \quad \forall x\in\Omega=]0,L[\\ u(0)=T_0 \\ u(L)=T_0 \end{cases}$$

This problem therefore serves as a model for problems of the following type:

$$\begin{cases} -u''(x)+c(x)u(x) &= f(x) \quad \forall x\in\Omega=]0,1[\\ u(0) &= \alpha \\ u(L) &= \beta \end{cases} \qquad [4.22]$$

In the model problem [4.22], f and c are known functions that are continuous on $[0,1]$, and $c(x) \geq 0$ for all x in $[0,1]$. We want to find a function $u(x)$, $x \in [0,1]$ that solves [4.22].

First, we partition the interval $[0,1]$ into equal parts using the points $x_i = ih$, $i = 0, \ldots, N+1$. This partition is said to be a uniform mesh on $[0,1]$, the points x_i, $i = 0, \ldots, N+1$ are said to be the nodes of the mesh, and $h = \frac{1}{N+1}$ is said to be the size of the mesh. As the number $(N+2)$ of nodes of the mesh tends to infinity, h tends to 0.

Write $u_i = u(x_i)$. Then, $u_0 = \alpha$ and $u_{N+1} = \beta$ satisfy the boundary conditions. We still need to find u_i, $i = 1, \ldots, N$. The differential equation of this problem may be written for each of the x_i as follows:

$$-u''(x_i) + c(x_i)u(x_i) = f(x_i) \quad i = 0, \ldots, N.$$

Since the functions c and f are known, we can calculate $c(x_i) = c_i$ and $f(x_i) = f_i$. We saw earlier that we can approximate the second derivative of a function at a given point using the formula:

$$\begin{cases} u''(x_i) = \dfrac{u(x_i+h) - 2u(x_i) + u(x_i-h)}{h^2} - \dfrac{h^2}{12}u^{(4)}(\xi),\ \xi \in [x_i - h, x_i + h] \\ u''(x_i) = \dfrac{u_{i+1} - 2u_i + u_{i-1}}{h^2} - \dfrac{h^2}{12}u^{(4)}(\xi),\ \xi \in [x_{i-1}, x_{i+1}]. \end{cases}$$

It is easy to see that this implies:

$$\max_{1 \leq i \leq N} \left| -u''(x_i) - \left(\frac{-u_{i+1} + 2u_i - u_{i-1}}{h^2} \right) \right| \leq \frac{h^2}{12} \sup_{x \in [0,1]} |u^{(4)}(x)|. \qquad [4.23]$$

Therefore, we deduce that there exist $\xi_i \in [x_{i-1}, x_{i+1}]$, $i = 1, \ldots, N$ such that:

$$\frac{-u_2 + 2u_1 - u_0}{h^2} + c_1 u_1 = f_1 - \frac{h^2}{12}u^{(4)}(\xi_1)$$

$$\frac{-u_3 + 2u_2 - u_1}{h^2} + c_2 u_2 = f_2 - \frac{h^2}{12}u^{(4)}(\xi_2)$$

$$\cdots$$

$$\frac{-u_{i+1} + 2u_i - u_{i-1}}{h^2} + c_i u_i = f_i - \frac{h^2}{12}u^{(4)}(\xi_i)$$

$$\cdots$$

$$\frac{-u_{N+1} + 2u_N - u_{N-1}}{h^2} + c_N u_N = f_N - \frac{h^2}{12}u^{(4)}(\xi_N).$$

This leads to the following system to be solved:

$$\frac{1}{h^2}\begin{pmatrix} 2+c_1h^2 & -1 & 0 & 0 & \dots & 0 \\ -1 & 2+c_2h^2 & -1 & 0 & \dots & \vdots \\ 0 & -1 & 2+c_3h^2 & -1 & \ddots & \vdots \\ \vdots & 0 & -1 & \ddots & \ddots & 0 \\ \vdots & & \ddots & \ddots & \ddots & -1 \\ 0 & \dots & \dots & 0 & -1 & 2+c_Nh^2 \end{pmatrix}\begin{pmatrix} u_1 \\ u_2 \\ \vdots \\ \vdots \\ \vdots \\ u_N \end{pmatrix}$$

$$= \begin{pmatrix} f_1 + \frac{\alpha}{h^2} \\ f_2 \\ \vdots \\ \vdots \\ \vdots \\ f_N + \frac{\beta}{h^2} \end{pmatrix} - \frac{h^2}{12}\begin{pmatrix} \epsilon_1 \\ \epsilon_2 \\ \vdots \\ \vdots \\ \vdots \\ \epsilon_N \end{pmatrix},$$

which is of the form $A_hU_h = B + E_h$, where A_h is a symmetric tridiagonal matrix and E_h is a vector such that:

$$\|E_h\|_\infty \leq \frac{h^2}{12} \sup_{x\in[0,1]} |u^{(4)}(x)|,$$

which therefore tends to 0 at the same rate as h^2.

It can be shown that the matrix A_h is positive definite, and so is invertible. Consider the following linear system:

$$A_hU_h = B, \qquad [4.24]$$

and write $U_h = (u_h(x_i))_{1\leq i\leq N}$ for the unique vector solution. The following result holds:

THEOREM 4.7.– *If $u \in C^4([0,1])$ is a solution of [4.22] and u_h is a solution of system [4.24], then*

$$\|u - u_h\|_\infty = \max_{1\leq i\leq N} |u(x_i) - u_h(x_i)| \leq \frac{h^2}{96} \sup_{x\in[0,1]} |u^{(4)}(x)|.$$

Therefore, as h becomes smaller (as the number of points in the mesh increases), the approximation of the exact solution of the problem improves, provided that the exact solution is regular.

4.7. Using *Matlab*

We can calculate derivatives in *Matlab* using the symbolic computation ToolBox. First, we need to declare symbolic variables using the *syms* command:

```
>> syms a x n
>> diff(x^2)

ans =
     2*x

>> diff(sin(x)*exp(-a*x^2))

ans =
     cos(x)*exp(-a*x^2)-2*sin(x)*a*x*exp(-a*x^2)
```

We can calculate the second derivative of a function as follows:

```
>> diff(x^n,2)

ans =
      x^n*n^2/x^2 - x^n*n/x^2
>> factor(ans)
     x^n*n*(n-1)/x^2
```

In general, the n-th derivative of a function $f(x)$ is returned by the command *diff(f(x),n)*. For example:

```
>> diff(x^4*exp(x),3)

ans =
     24*x*exp(x)+36*x^2*exp(x)+12*x^3*exp(x)+x^4*exp(x)
>> factor(ans)
     x*exp(x)*(24+36*x+12*x^2+x^3)
```

5

Numerical Integration

5.1. Introduction

Given a function f that is integrable on $[a, b]$, numerical integration finds an approximate value of $\int_a^b f(x)dx$ using a quadrature formula:

$$\int_a^b f(x)dx = \underbrace{\sum_{i=0}^{n} A_i f(x_i)}_{\text{quadratureformula}} + \underbrace{R_n(f).}_{\text{quadratureerror}} \qquad [5.1]$$

The $x_i \in [a, b]$ are said to be the nodes of the formula and the A_i are known as the weights. The classical approach to deriving the quadrature formula is as follows. Choose $(n + 1)$ distinct points $x_i \in [a, b]$ such that:

$$a = x_0 < x_1 < x_2 < \cdots < x_{n-1} < x_n = b$$

$$h = \frac{b-a}{n} \text{ and } x_i = a + ih, \; i = 0, 1, \ldots, n.$$

This allows us to rewrite the integral:

$$\int_a^b f(x)dx = \sum_{i=0}^{n-1} \int_{x_i}^{x_{i+1}} f(x)dx. \qquad [5.2]$$

We then approximate each of these integrals $\int_{x_i}^{x_{i+1}} f(x)dx$ below using quadrature formulas.

The first step is to replace f by p (the interpolating polynomial on the points x_i). Thus:

$$f(x) = p(x) + E_n(f)$$
$$= \sum_{i=0}^{n-1} l_i(x) f(x_i) + E_n(f),$$

where the $l_i(x)$ are the Lagrange polynomials and $E_n(f)$ is the interpolation error.

The quadrature formula can now be obtained by integrating; the error of the quadrature can be deduced from the formula of averages:

$$\int_a^b f(x)dx = \sum_{i=0}^{n} A_i f(x_i) + R_n(f)$$

$$A_i = \int_a^b l_i(x)dx \quad \text{and} \quad R_n(f) = \int_a^b E_n(f)dx.$$

5.2. Rectangle method

This method replaces the integrand f by a piecewise constant function on each subinterval $[x_i, x_{i+1}]$. We can choose to use either left rectangles, i.e. $f(x_i)$ for $x \in [x_i, x_{i+1}]$ (see Figure 5.1), giving:

$$\int_a^b f(x)dx \simeq \sum_{i=0}^{n-1} (x_{i+1} - x_i) f(x_i),$$

or right rectangles, i.e. $f(x_{i+1})$ for $x \in [x_i, x_{i+1}]$, giving:

$$\int_a^b f(x)dx \simeq \sum_{i=0}^{n-1} (x_{i+1} - x_i) f(x_{i+1}).$$

The rectangle method is of order 0. If the first derivative of f is bounded by a constant M, the error satisfies:

$$\left| \int_a^b f(x)dx - h \sum_{i=0}^{n-1} f(a + ih) \right| \leq \frac{1}{2} \frac{(b-a)^2}{n} M.$$

5.3. Trapezoidal rule

Choose $x_i = a + ih$, where $h = \frac{b-a}{n}$. On each subinterval $[x_i, x_{i+1}]$, we will replace the function f by a function that is linear on that subinterval. Thus:

$$\int_{x_i}^{x_{i+1}} f(x)dx = \frac{h}{2}(f(x_i) + f(x_{i+1})) + R_i(f),$$

and, assuming that f is twice continuously differentiable on $[a, b]$:

$$\begin{aligned} R_i(f) &= \frac{h^3}{2} f''(\xi_i) \int_0^1 t(t-1)dt \quad \xi_i \in [x_i, x_{i+1}] \\ &= -\frac{h^3}{12} f''(\xi_i). \end{aligned}$$

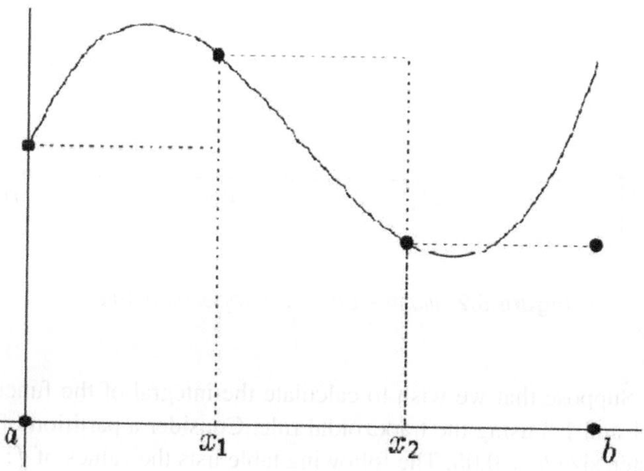

Figure 5.1. *Illustration of the rectangle method*

Therefore:

$$\begin{aligned} \int_a^b f(x)dx &= \sum_{i=0}^{n-1} \int_{x_i}^{x_{i+1}} f(x)dx \\ &= \frac{h}{2} \sum_{i=0}^{n-1} (f(x_i) + f(x_{i+1})) - \frac{h^3}{12} \sum_{i=0}^{n-1} f''(\xi_i), \end{aligned}$$

which implies the trapezoidal rule (see Figure 5.2):

$$\int_a^b f(x)dx = \frac{h}{2}(f(a) + 2f(a+h) + \cdots + 2f(a+(n-1)h) + f(b)) + R(f),$$

where $R(f) = -\frac{(b-a)^3}{12n^2} f''(\xi)$ and $\xi \in]a, b[$. It is clear that $\lim_{n \to \infty} R(f) = 0$.

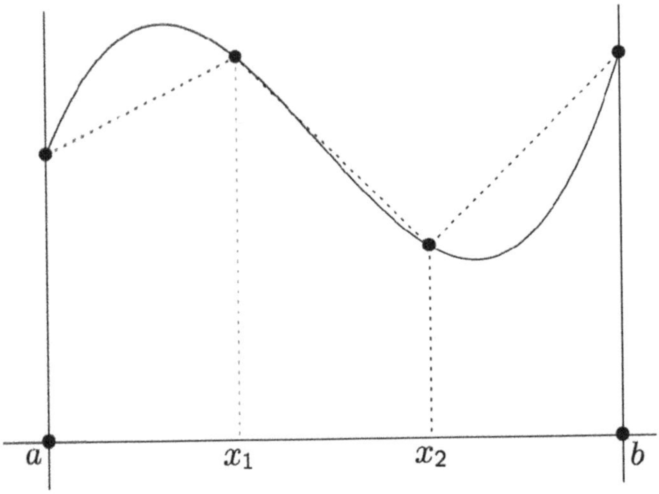

Figure 5.2. *Illustration of the trapezoidal rule*

EXAMPLE.– Suppose that we wish to calculate the integral of the function $f(x) = \sqrt{x}$ between 1 and 1.3 using the trapezoidal rule. Consider a partition of the interval $[1, 1.3]$ with step size $h = 0.05$. The following table lists the values of f:

x	$\sqrt{x}$
$x_0 = 1$	1
$x_1 = 1.05$	1.02470
$x_2 = 1.10$	1.04881
$x_3 = 1.15$	1.07238
$x_4 = 1.20$	1.09544
$x_5 = 1.25$	1.11803
$x_6 = 1.30$	1.14017

Therefore, by the trapezoidal rule:

$$\int_{1.0}^{1.3} \sqrt{x}dx \simeq \frac{0.05}{2}[f(x_0) + 2(f(x_1) + \cdots + f(x_5)) + f(x_6)]$$
$$= 0.32147.$$

The exact value obtained by evaluating the primitive is 0.32149. The effective error in this example is 2×10^{-5}.

5.4. Simpson's rule

Simpson's rule uses a three-point Newton-Côtes formula on each subinterval, which is written as a function of the midpoint of each subinterval. For simplicity, we shall adopt the convention of labeling the endpoints of the subintervals in pairs, reserving the odd indices for the midpoints of each interval. In other words, set $x_i = a + ih$ for $i = 0, \ldots, n$ and $h = (b - a)/h$ for even n (see Figure 5.3):

$$\int_a^b f(x)dx = \int_a^{x_2} f(x)dx + \int_{x_2}^{x_4} f(x)dx + \cdots + \int_{x_{n-2}}^b f(x)dx.$$

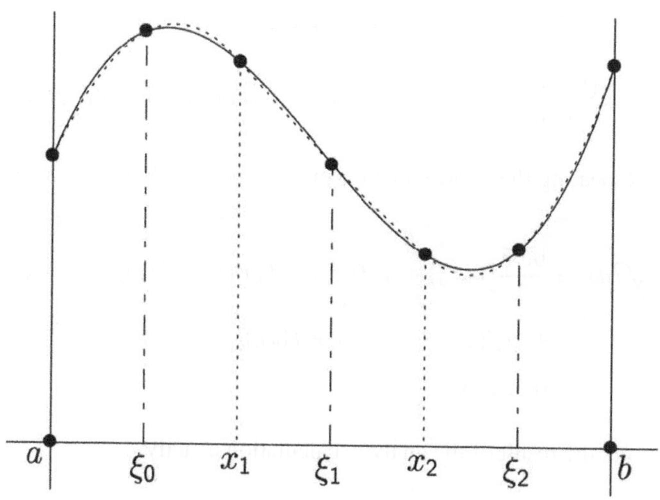

Figure 5.3. *Illustration of Simpson's rule*

Then:

$$\int_{x_{2i}}^{x_{2i+2}} f(x)dx = \frac{h}{3}(f(x_{2i}) + 4f(x_{2i+1}) + f(x_{2i+2})) + R_i(f).$$

If we assume that f is four times continuously differentiable on $[a, b]$:

$$R_i(f) = \frac{h^5}{4!} f^{(4)}(\xi_i) \int_0^2 t^2(t-1)(t-2)dt \quad \xi \in [x_{2i}, x_{2i+2}]$$

$$= -\frac{h^5}{90} f^{(4)}(\xi_i).$$

Therefore:

$$\int_a^b f(x)dx = \sum_{i=0}^{\frac{n}{2}-1} \int_{x_{2i}}^{x_{2i+2}} f(x)dx$$

$$= \frac{h}{3} \sum_{i=0}^{\frac{n}{2}-1} (f(x_{2i}) + 4f(x_{2i+1}) + f(x_{2i+2})) - \frac{h^5}{90} \sum_{i=0}^{\frac{n}{2}-1} f^{(4)}(\xi_i),$$

from which we deduce Simpson's rule:

$$\int_a^b f(x)dx = \frac{h}{3}(f(a) + 4f(a+h) + 2f(a+2h) + \cdots + 2f(a+(n-2)h)$$
$$+4f(a+(n-1)h) + f(b)) + R(f),$$

where $R(f) = -\frac{(b-a)^5}{180n^4} f^{(4)}(\xi)$ and $\xi \in]a,b[$. It is clear that: $\lim_{n\to\infty} R(f) = 0$.

EXAMPLE.– Repeating the approximation in Example 5.3, this time with Simpson's rule:

$$\int_{1.0}^{1.3} \sqrt{x}dx \simeq \frac{0.05}{3}[f(x_0) + 4(f(x_1) + f(x_3) + f(x_5))$$
$$+ 2(f(x_2) + f(x_4)) + f(x_6)]$$
$$= 0.32149.$$

This recovers the result of the analytic calculation exactly.

EXAMPLE.– Let $f \in C([-1,1], \mathbb{R})$. The quadratic polynomial p that interpolates f on the points $t_0 = -1$, $t_1 = 0$ and $t_2 = 1$ may be written as:

$$p(t) = f(-1)l_0(t) + f(0)l_1(t) + f(1)l_2(t),$$

where $\{l_0, l_1, l_2\}$ is the Lagrange basis of $\mathbb{P}_2$ for the points $t_0 = -1$, $t_1 = 0$, and $t_2 = 1$, and:

$$l_0(t) = \frac{t^2}{2} - \frac{t}{2}, \; l_1(t) = 1 - t^2, \text{ and } l_2(t) = \frac{t^2}{2} + \frac{t}{2}.$$

A straightforward computation shows that:

$$\int_{-1}^1 p(t)dt = f(-1)\int_{-1}^1 l_0(t)dt + f(0)\int_{-1}^1 l_1(t)dt + f(1)\int_{-1}^1 l_2(t)dt.$$

Now:

$$\int_{-1}^1 l_0(t)dt = \frac{1}{3} \int_{-1}^1 l_1(t)dt = \frac{4}{3} \int_{-1}^1 l_2(t)dt = \frac{1}{3},$$

so:

$$\int_{-1}^{1} p(t) = \frac{1}{3}[f(-1) + 4f(0) + f(1)] := J(f).$$

Thus, $J(f)$ gives a natural approximation of $\int_{-1}^{1} f(t)dt$.

This formula $J(f)$ is known as Simpson's quadrature formula. It integrates cubic polynomials exactly, in the sense that:

$$\int_{-1}^{1} q(t)dt = J(q) \quad \forall q \in \mathbb{P}_3.$$

Equivalently:

$$\int_{-1}^{1} t^n dt = J(t^n) \quad \forall n = 0, 1, 2, 3.$$

It is easy to verify that $\int_{-1}^{1} t^4 \neq J(t^4)$.

EXAMPLE.– Suppose that $f(x) = \frac{1}{x}$ and write p_2 for the interpolating polynomial on the three points $x_0 = 1$, $x_1 = 2$, and $x_3 = 3$, with step size $h = 1$. Then:

$$\begin{aligned} f(x) &= p_2(x) + E(x) \\ &= \frac{x^2 - 6x + 11}{6} + E(x), \end{aligned}$$

where $E(x) = \frac{1}{3!}\pi(x)f^{(3)}$ (for $\xi_x \in [1, 3]$) and $\pi(x) = (x - x_0)(x - x_1)(x - x_2)$.

The three points are an equal distance $h = 1$ apart, so, by Simpson's rule:

$$\begin{aligned} \int_{x_1}^{x_3} p_2(x)dx &= \int_{1}^{3} p_2(x)dx \\ &= \frac{h}{3}(f(x_0) + 4f(x_1) + f(x_2)) \\ &= \frac{1}{3}(f(1) + 4f(2) + f(3)) \\ &= 1 + \frac{1}{9}. \end{aligned}$$

The term $\pi(x)$ changes sign on $[1, 3]$; it is positive on $[1, 2]$ and negative on $[2, 3]$. Therefore:

$$\left| \int_{1}^{3} E(x)dx \right| \leq \int_{1}^{2} \pi(x)dx - \int_{2}^{3} \pi(x)dx.$$

Now, since $\pi(x) = x^3 - 6x^2 + 11x - 6$:

$$\int_1^2 \pi(x)dx - \int_2^3 \pi(x)dx = \frac{1}{2} = 5 \times 10^{-1}.$$

This only gives a rough estimate of the error. Alternatively, we can write $E(x)$ as:

$$E(x) = \pi(x)f(x, 1, 2, 3) = (x^3 - 6x^2 + 11x - 6) \times \left(-\frac{1}{6x}\right),$$

which implies that:

$$\begin{aligned}\int_1^3 E(x)dx &= -\frac{10}{9} + \ln 3\\ &\simeq -1.25 \times 10^{-2},\end{aligned}$$

hence:

$$\left|\int_1^3 E(x)dx\right| \leq 1.25 \times 10^{-2}.$$

5.5. Hermite's rule

Let f be a function with class C^4 on the interval $[a, b]$, and suppose that P is its Hermite interpolating polynomial on this interval, which satisfies:

$$P(a) = f(a) \quad P'(a) = f'(a) \quad P(b) = f(b) \quad P'(b) = f'(b).$$

We can write this polynomial as:

$$\begin{aligned}P(x) = f(a)&\left[\frac{(x-b)^2}{(a-b)^2} + 2\frac{(x-a)(x-b)^2}{(b-a)^3}\right] + f'(a)\left[\frac{(x-a)(x-b)^2}{(b-a)^2}\right]\\ &+f(b)\left[\frac{(x-a)^2}{(b-a)^2} - 2\frac{(x-a)^2(x-b)}{(b-a)^3}\right] + f'(b)\left[\frac{(x-a)^2(x-b)}{(b-a)^2}\right],\end{aligned}$$

and an upper bound for the error is given by:

$$|f(x) - P(x)| \leq \frac{M_4}{4!}(x-a)^2(x-b)^2,$$

where $M_4 = \sup_{t\in[a,b]} |f^{(4)}(t)|$.

Next, write $I(f) = \int_a^b f(x)dx$. The value of $\tilde{I}(f) = \int_a^b P(x)dx$ gives an approximation of $I(f)$, and can be written as:

$$\tilde{I}(f) = \frac{(b-a)(f(a) + f(b))}{2} + \frac{(b-a)^2(f'(a) - f'(b))}{12}.$$

The error $E = |\tilde{I}(f) - I(f)|$ satisfies:

$$E = |\int_a^b (f(x) - P(x))dx| \leq \int_a^b |(f(x) - P(x))|dx$$
$$\leq \int_a^b \frac{M_4}{24}(x-a)^2(x-b)^2 dx \leq \frac{(b-a)^5 M_4}{720}.$$

This method can be improved by applying it multiple times after dividing the interval $[a, b]$ into n intervals of equal width $h = \dfrac{b-a}{n}$. By writing $x_i = a + ih$ for $0 \leq i \leq n$, the following integral approximation formula holds:

$$I(f) \simeq \tilde{I}_n(f) = \frac{h}{2}(f(x_0) + 2f(x_1) + 2f(x_2) + \ldots$$
$$+ 2f(x_{n-1}) + 2f(x_n)) + \frac{h^2}{12}(f'(a) - f'(b)).$$

This is called Hermite's rule, or the corrected trapezoidal rule.

This allows us to replace f with the piecewise Hermite polynomial approximation P_n and thus approximate $I(f)$ by:

$$\tilde{I}_n(f) = \int_a^b P_n(x)dx = \sum_{i=1}^{n} \int_{x_{i-1}}^{x_i} P_n(x)dx.$$

On each interval $[x_{i-1}, x_i]$:

$$\int_a^b P_n(x)dx = \frac{h}{2}(f(x_i) + f(x_{i-1})) + \frac{h^2}{12}(f'(x_{i-1}) - f'(x_i)).$$

By summing all of these terms together, we find:

$$\tilde{I}(f) = \frac{(b-a)(f(a) + f(b))}{2} + \frac{(b-a)^2(f'(a) - f'(b))}{12}.$$

The error is $E_n = |\tilde{I}_n(f) - I(f)| \leq \dfrac{(b-a)^5}{720n^4} M_4$.

REMARK.– This method is similar to Simpson's rule, but requires precisely two derivative computations, $f'(a)$ and $f'(b)$, regardless of the number of intervals.

5.6. Newton–Côtes rules

The Newton–Côtes rules generalize the trapezoidal rule and Simpson's rule: the function f is approximated by a polynomial of degree n. The integral is evaluated using the expression:

$$\int_a^b f(x)dx \simeq a_0 f(x_0) + a_1 f(x_1) + \cdots + a_n f(x_n).$$

We can deduce the coefficients a_j from the condition that the relation should be exact whenever f is a polynomial of degree n or lower. By choosing $f(x) = x^k$ for each $k = 0, 1, \ldots, n$, we obtain the following linear system:

$$\begin{cases} a_0 + a_1 + \cdots + a_n & = b - a \\ a_0x_0 + a_1x_1 + \cdots + a_nx_n & = \frac{b^2-a^2}{2} \\ \vdots & \vdots \quad \vdots \\ a_0x_0^n + a_1x_1^n + \cdots + a_nx_n^n & = \frac{b^n-a^n}{n+1} \end{cases}.$$

The determinant of this system is a Vandermonde determinant with value $(x_0 - x_1)(x_1 - x_2)\ldots(x_n - x_0)$. When the points are equally spaced, we recover the Newton–Côtes formulas.

For $n = 1$, we recover the trapezoidal rule, and, for $n = 2$, we recover Simpson's rule.

For $n = 3$:

$$\int_{x_0}^{x_1} f(x)dx \simeq \frac{3h}{8}(f(x_0) + 3f(x_1) + 3f(x_2) + f(x_3)).$$

For $n = 4$ (Villarceau's rule):

$$\int_{x_0}^{x_1} f(x)dx \simeq \frac{2h}{45}(7f(x_0) + 32f(x_1) + 12f(x_2) + 32f(x_3) + 7f(x_4)).$$

5.7. Gauss–Legendre method

5.7.1. *Problem statement*

Let $[a, b]$ be an interval of $\mathbb{R}$ and suppose that g is a real function. We want to compute:

$$\int_a^b g(x)dx = \sum_{k=0}^{n} \beta_k g(y_k) + R_n(f).$$

The Gauss–Legendre method is as follows:

1) rewrite $\int_a^b g(x)dx = K\int_{-1}^{1} f(t)dt$ with the substitution

$$x = \frac{1}{2}(b-a)t + \frac{1}{2}(a+b);$$

2) choose $(x_i)_{i=0,\ldots,n}$ and $(\alpha_i)_{i=0,\ldots,n}$ such that:

$$\int_{-1}^{1} f(t)dt = \sum_{k=0}^{n} \alpha_k f(x_k) + R_n(f)$$

is exact for every polynomial in $\mathcal{P}_{2n+1}$.

EXAMPLE.– Let $x_1, x_2 \in [-1, 1]$, with $x_1 < x_2$ and $\omega_1, \omega_2 \in \mathbb{R}$, and suppose that $f \in \mathcal{C}([-1,1], \mathbb{R})$.

Consider the following quadrature formula on two points:

$$J(f) = \omega_1 f(x_1) + \omega_2 f(x_2). \quad [5.3]$$

We need to find x_1, x_2, ω_1, and ω_2 that maximize the degree of the polynomials for which this quadrature formula is exact.

The quadrature formula [5.3] is exact for polynomials of degree 0, so:

$$J(1) = \int_{-1}^{1} 1dt \Leftrightarrow \omega_1 + \omega_2 = 2.$$

Therefore, $\omega_2 = 2 - \omega_1$.

The quadrature formula [5.3] is exact for polynomials of degree 1, so:

$$J(t) = \int_{-1}^{1} tdt.$$

Therefore, $\omega_1 x_1 + \omega_2 x_2 = 0$.

The quadrature formula [5.3] is exact for polynomials of degree 2, so:

$$J(t^2) = \int_{-1}^{1} t^2 dt,$$

which implies that $\omega_1 x_1^2 + \omega_2 x_2^2 = \frac{2}{3}$. This leads us to the system:

$$\omega_1 = 2 - \omega_2 \quad [5.4]$$

$$\omega_1 x_1 = -\omega_2 x_2 \quad [5.5]$$

$$\omega_1 x_1^2 + \omega_2 x_2^2 = \frac{2}{3}. \quad [5.6]$$

By substituting equation [5.5] into equation [5.6], we find that:

$$\frac{2}{3} = \omega_2 x_2 [x_2 - x_1] \Rightarrow \omega_2 \neq 0,\ x_2 \neq 0,$$

$$\frac{2}{3} = \omega_1 x_1 [x_1 - x_2] \Rightarrow \omega_1 \neq 0,\ x_1 \neq 0.$$

This quadrature formula is exact for polynomials of degree 3. Therefore, $\omega_1 x_1^3 + \omega_2 x_2^3 = 0$, and so:

$$\omega_1 x_1 (x_1 - x_2)(x_1 + x_2) = 0.$$

Therefore, $x_1 = -x_2$, which implies that $\omega_1 = \omega_2$ by equation [5.5]. Equation [5.4] then implies that $\omega_1 = \omega_2 = 1$. By equation [5.6]:

$$2x_1^2 = \frac{2}{3}.$$

Therefore:

$$x_1 = -\frac{1}{\sqrt{3}} \quad \text{and} \quad x_2 = \frac{1}{\sqrt{3}}.$$

The points x_1 and x_2 are known as the Gaussian points (see below).

5.7.2. *Legendre polynomials*

DEFINITION.– *The Legendre polynomials are defined by the formula*

$$L_n(x) = \frac{1}{2^n n!}\frac{d^n}{dx^n}((x^2-1)^n).$$

For example, for $x \in \mathbb{R}$,

$$L_0(x) = 1,\ L_1(x) = x,\ L_2(x) = \frac{3x^2-1}{2}.$$

The Legendre polynomials L_0, L_1, L_2, etc. satisfy the following properties:

1) $I_{n,m} = \int_{-1}^{1} L_n(x)L_m(x)dx = \begin{cases} 0 & n \neq m \text{ (orthogonality)}; \\ \frac{2}{2n+1} & n = m \end{cases}$

2) the set $\{L_0, L_1, \ldots, L_n\}$ is a basis of $\mathcal{P}_n$;

3) the polynomial L_n has n distinct roots, all of which are contained in the open interval $(-1,+1)$. These roots are called the Gaussian points;

4) the function $\psi(u,x) = \sum_{i=0}^{\infty} L_i(x)u^i$ is called the generating function of the (L_n), and has the expression:

$$\psi(u,x) = (1-2xu+u^2)^{\frac{-1}{2}};$$

5) $L_n(x) = \dfrac{2n-1}{n}xL_{n-1}(x) - \dfrac{n-1}{n}L_{n-2}(x)$;

6) $\int_1^1 [L_n(x)]^2 dx = \dfrac{2}{2n+1}.$

We will justify some parts of this theorem.

1) First, we show that, $\forall n \geq 1, \forall l < n$:

$$\frac{d^l}{dx^l}(x^2-1)^n = 0 \text{ for } x = \pm 1. \qquad [5.7]$$

We argue by induction on n.

For $n = 1$, the fact that $l < n$ implies that $l = 0$, so $(x^2 - 1) = 0$ for $x = \pm 1$.

Suppose now that the following relation holds up to order n:

$$\forall l < n \quad \frac{d^l}{dx^l}(x^2 - 1)^n = 0 \text{ for } x = \pm 1.$$

We will now prove that it also holds for the case $n + 1$.

– If $l = 0$, then:

$$\frac{d^0}{dx^0}(x^2 - 1)^{n+1} = (x^2 - 1)^{n+1} = 0 \text{ for } x = \pm 1.$$

– If $l = 1, \ldots, n + 1$, then we can write l as $l = l' + 1$, where $l' = 0, \ldots, n$. This implies that:

$$\begin{aligned}\frac{d^l}{dx^l}(x^2 - 1)^{n+1} &= \frac{d^{l'}}{dx^{l'}}\left(\frac{d}{dx}(x^2 - 1)^{n+1}\right) \\ &= \frac{d^{l'}}{dx^{l'}}((n + 1)(x^2 - 1)^n 2x) \\ &= \sum_{r=0}^{l'} C_{l'}^r \frac{d^r}{dx^r}(x^2 - 1)^n (2x)^{l'-r} \\ &= C_{l'}^{l'}(2x)\frac{d^{l'}}{dx^{l'}}(x^2 - 1)^n + C_{l'}^{l'-1} 2 \frac{d^{l'-1}}{dx^{l'-1}}(x^2 - 1)^n.\end{aligned}$$

Finally:

$$\frac{d^{l'}}{dx^{l'}}(x^2 - 1)^n = \frac{d^{l'-1}}{dx^{l'-1}}(x^2 - 1)^n = 0 \text{ for } x = \pm 1,$$

which gives the desired result. Taking note of the equality:

$$\int_{-1}^{1} x^k L_n(x) dx = \frac{1}{2^n n!}\int_{-1}^{1} \frac{d^n}{dx^n}(x^2 - 1)^n \times x^k dx,$$

we can also show that:

$$\int_{-1}^{1} x^k \frac{d^n}{dx^n}(x^2-1)^n dx = \left[x^k \frac{d^{n-1}}{dx^{n-1}}(x^2-1)^n\right]_{-1}^{1}$$

$$- \int_{-1}^{1} kx^{k-1} \frac{d^{n-1}}{dx^{n-1}}(x^2-1)^n dx$$

$$= (-1)^k k! \int_{-1}^{1} \frac{d^{n-k}}{dx^{n-k}}(x^2-1)^n dx$$

$$= (-1)^k k! \left[\frac{d^{n-k+1}}{dx^{n-k+1}}(x^2-1)^n dx\right]_{-1}^{1} = 0.$$

2) Suppose that $m < n$ (if not, reverse them).

$$\int_{-1}^{1} x^k P_n(x) dx = 0 \quad \forall k = 0, 1, \ldots, n-1,$$

so, for each $k = 0, \ldots, m$, the coefficient a_k satisfies:

$$\sum_{k=0}^{m} a_k \int_{-1}^{1} x^k L_n(x) dx = 0 \Rightarrow \int_{-1}^{1} (\sum_{k=0}^{m} a_k x^k) L_n(x) dx = 0.$$

As $L_m(x)$ is a polynomial of degree m, there exist coefficients a_k such that $L_m(x) = \sum_{k=0}^{m} a_k x^k$, which implies that:

$$\int_{-1}^{1} L_m(x) L_n(x) dx = 0 \quad \forall m \neq n.$$

3) Let $\mathbb{R}_n[X]$ be the vector space of polynomials of degree less than or equal to n. For each k, L_k is a polynomial of degree k. Note that $L_0(x) = 1$ and $L_1(x) = 2x$. Suppose that the property holds up to order n. We will show that it also holds for order $n+1$:

$$L_{n+1}(x) = \frac{1}{2^{n+1}(n+1)!} \frac{d^{n+1}}{dx^{n+1}}(x^2-1)^{n+1}$$

$$= c_{n+1} \left[\frac{d^n}{dx^n}\left(\frac{d}{dx}(x^2-1)^{n+1}\right)\right]$$

$$= (n+1)c_{n+1}(2x)\frac{d^n}{dx^n}(x^2-1)^n,$$

so L_n is a polynomial of degree $n+1$.

It can easily be shown that $(L_k(x))_{0\leq k\leq n}$ is a free family in $\mathbb{R}_n[X]$, which means that it is a basis. Let $P \in \mathbb{R}_n[X]$. Then, $P = \sum_{k=0}^{n} a_k L_k$, where:

$$a_k = \frac{\int_{-1}^{1} P(x)L_k(x)dx}{\int_{-1}^{1}(L_k(x))^2 dx}.$$

Therefore, every polynomial of degree n has a unique expression in terms of the polynomials L_k for $k = 0, 1, \ldots, n$.

4) First, we show that:

$$\int_{-1}^{1} x^n L_n(x)dx = \frac{2^{n+1}(n!)^2}{(2n+1)!}.$$

Setting $k = n$ in the result that we proved earlier in (1):

$$\begin{aligned}\int_{-1}^{1} x^n \frac{d^n}{dx^n}(x^2-1)^n dx &= (-1)^n n! \int_{-1}^{1} (x^2-1)^n dx \\ &= 2n! \int_{0}^{1} (1-x^2)^n dx \\ &= 2(n!) \int_{0}^{\frac{\pi}{2}} \cos^{2n+1} t dt.\end{aligned}$$

Note also that:

$$\begin{aligned}\int_{0}^{\frac{\pi}{2}} \cos^{2n+1} t dt &= \left[\frac{\cos^{2n} t \sin t}{2n+1}\right]_0^{\frac{\pi}{2}} + \frac{2n}{2n+1}\int_{0}^{\frac{\pi}{2}} \cos^{2n-1} t dt \\ &= \frac{2n(2n-1)\ldots 2}{(2n+1)(2n-1)\ldots 3}\int_{0}^{\frac{\pi}{2}} \cos t dt.\end{aligned}$$

Therefore:

$$\int_{-1}^{1} x^n \frac{d^n}{dx^n}(x^2-1)^n dx = 2n! \frac{2n(2n-1)\ldots 2}{(2n+1)(2n-1)\ldots 3}.$$

Multiplying the numerator and the denominator by $2n(2n-2)\dots 2$:

$$\int_{-1}^{1} x^n L_n(x)dx = \frac{1}{2^n n!} \times 2n! \times \frac{2^n n!\, 2^n n!}{(2n+1)!}$$
$$= \frac{2^{n+1}(n!)^2}{(2n+1)!}.$$

We will show that:

$$L_n(x) = \frac{1}{2^n n!}\frac{(2n)!}{n!}x^n + \cdots$$

We will show by induction that:

$$\frac{d^n}{dx^n}(x^2-1)^n = \frac{(2n)!}{n!}x^n + \cdots$$

– $n = 0$, $P_0(x) = 1$.

– $n = 1$, $\frac{d^1}{dx^1}(x^2-1) = 2x = \frac{2\times 1!}{1!}x^1$

Suppose that the claim holds up to order n, then:

$$\frac{d^{n+1}}{dx^{n+1}}(x^2-1)^{n+1} = \frac{d^n}{dx^n}\left(\frac{d}{dx}(x^2-1)^{n+1}\right)$$
$$= 2(n+1)\frac{d^n}{dx^n}(x^2-1)^n.x$$
$$= 2(n+1)\sum_{k=0}^{n} \mathrm{C}_n^k((x^2-1)^n)^{(k)}x^{(n-k)}$$
$$= 2(n+1)\left[\mathrm{C}_n^n x\frac{d^n}{dx^n}(x^2-1)^n + \mathrm{C}_n^{n-1}\frac{d^{n-1}}{dx^{n-1}(x^2-1)^n}\right]$$
$$= 2(n+1)\left[\frac{(2n)!}{n!} + \frac{n(2n)!}{(n+1)!}\right]x^{n+1} + \cdots$$
$$= 2(n+1)\left[1 + \frac{n}{n+1}\right]\frac{(2n)!}{n!}x^{n+1} + \cdots$$
$$= \frac{(2n+2)!}{(n+1)!}x^{n+1} + \cdots$$

Therefore, by induction:

$$\frac{d^n}{dx^n}(x^2-1)^n = \frac{(2n)!}{n!}x^n + \cdots$$

Hence:

$$\begin{aligned}\int_{-1}^{1} L_n^2(x)dx &= \int_{-1}^{1}\left[\frac{1}{2^n n!}\frac{(2n)!}{n!}x^n + \cdots\right] L_n(x)dx\\ &= \frac{1}{2^n n!}\frac{(2n)!}{n!}\int_{-1}^{1} L_n(x)dx\\ &= \frac{1}{2^n n!}\frac{(2n)!}{n!}\frac{2^{n+1}(n!)^2}{(2n+1)!}\\ &= \frac{2}{2n+1}.\end{aligned}$$

5.7.3. *Choosing the* α_i *and* x_i *(*$i = 0, \ldots, n$*)*

$$\int_{-1}^{1} f(t)dt = \sum_{k=0}^{n} \alpha_k f(x_k) + R \qquad [5.8]$$

Formula [5.8] is exact for every polynomial $Q \in \mathcal{P}_{2n+1}$. As deg $L_{n+1} = n+1$, there exist P and $R \in \mathcal{P}_n$ such that $Q = PL_{n+1} + R$, where deg $R \leq n$ and deg $Q \leq n$:

$$[5.8] \text{ is exact on } \mathcal{P}_{2n+1} \iff \begin{cases} i) \ \forall P \in \mathcal{P}_n \ [5.8] \text{ is exact for } PL_{n+1} \\ ii) \qquad [5.8] \text{ is exact on } \mathcal{P}_n \end{cases}$$

5.7.3.1. *Choosing the* (x_i) $i = 0, \ldots, n$

Let $P \in \mathcal{P}_n$. The formula [5.8] is exact for PL_{n+1}. Therefore:

$$\int_{-1}^{1} P(x)L_{n+1}(x)dx = \sum_{i=0}^{n} \alpha_i P(x_i)L_{n+1}(x_i),$$

and so:

$$\sum_{i=0}^{n} \alpha_i P(x_i)L_{n+1}(x_i) = 0.$$

Thus, it suffices to ensure that:

$$L_{n+1}(x_i) = 0 \quad i = 0, \ldots, n, \qquad [5.9]$$

and so the x_i are the roots of L_{n+1}.

5.7.3.2. *Choosing the* (α_i) $i = 0, \ldots, n$

Formula [5.8] is exact on $\mathcal{P}_n$, and the x_i satisfy [5.9], so:

$$\int_{-1}^{1} x^k dx = \sum_{i=0}^{n} \alpha_i x_i^k. \qquad [5.10]$$

Equation [5.10] is a system of $(n+1)$ equations and $(n+1)$ unknowns, and therefore has a unique solution.

Alternatively, let:

$$\begin{cases} g_i(x) = \dfrac{L_{n+1}(x)}{x - x_i} & x \neq x_i \\ g_i(x_i) = L'_{n+1}(x_i) \end{cases}$$

then:

$$\int_{-1}^{1} g_i(x) dx = \sum_{l=0}^{n} \alpha_l g_i(x_l) = \alpha_i g_i(x_i) \quad i = 0, \ldots, n,$$

so:

$$\alpha_i = \frac{1}{L'_{n+1}(x_i)} \int_{-1}^{1} \frac{L_{n+1}(x)}{x - x_i} dx.$$

PROPOSITION.– *Let $x_0, x_1, \ldots, x_n$ be the roots of the Legendre polynomial L_{n+1} and define:*

$$\alpha_i = \frac{1}{L'_{n+1}(x_i)} \int_{-1}^{1} \frac{L_{n+1}(x)}{x - x_i} dx.$$

Then, the formula:

$$\int_{-1}^{1} f(t) dt = \sum_{i=0}^{n} \alpha_i f(x_i) + R_n(f)$$

is exact for every polynomial in $\mathcal{P}_{2n+1}$.

5.8. Using *Matlab*

5.8.1. Matlab *functions for numerical integration*

Matlab offers two functions “quad” and “quad8” for numerical integration. The “quad” function uses Simpson’s method, and the “quad8” function uses the Newton–Côtes formulas of order 8. These two functions may be used as follows:

– quad(“function_f”,a,b);

– quad(“function_f”,a,b,tol);

– quad("function_f",a,b,tol,trace).

The first form leaves the tolerance "tol" unspecified. This parameter defines the relative error E and is taken to be 0.001 by default. The integral quadrature computation is iterated until the tolerance is satisfied. If the third form is used with a non-zero "trace" value, a figure plotting the progression of the iterations of the algorithm is displayed on the screen. The "quad8" function is used in the same way as the "quad" function, and can similarly be called in any of the three ways shown above. For example:

```
>> quad('sqrt(1+exp(x))',0,2)

ans =

     0.40070
```

5.8.2. *Trapezoidal rule*

If the function f is known at the regular intervals ($x_i + x_{i-1} = h$), the trapezoidal rule can be programmed in *Matlab* as follows:

$$I = h * (sum(f) - 0.5 * (f(1) + f(length(f)))),$$

where:

$$sum(f) = f_1 + f_2 + f_3 + \cdots + f_n + f_{n+1}.$$

We can also write a program to calculate the integral I. One example of such a program ("trapez_v.m") is listed below:

```
function I=trapez_v(g,h) I=(sum(f)-(f(1)+f(length(f)))/2)*h;
```

Suppose, for example, that we wish to integrate the function $f(x) = x^2 + 2x + 1$ over the interval $[-10, 8]$, with a step size h of 1. In *Matlab* interactive mode (command mode), before running the program "trapez_v.m", we need to define the function and the bounds of integration:

```
>>x=-10:1:8;
>>f=x.^2+2*x-1;
>>h=1;
```

We then execute the script "trapez_v.m". This gives:

```
>>I=trapez_v(f,h)
I = 2265
```

The scripts "trapez_n.m" and "trapez_g.m" shown below can also be used to calculate the integral I:

```
>>I=trapez_n('function_f',a,b,n)
```

and:

```
>>I=trapez_g('function_f',a,b,n)
```

The code of "trapez_n.m" is:

```
function I=trapez_n(function_f,a,b,n)
h=(b-a)/n;
x=a+(0:n)*h;
f=feval(function_f,x);
I=trapez_v(f,h)
```

The code of "trapez_g.m" is:

```
function I=trapez_g(function_f,a,b,n)
n=n; hold off;
h=(b-a)/n;
x=a+(0:n)*h;
f=feval('function_f',x);
I=h/2*(f(1)+f(n+1));
if n>1
    I=I+sum(f(2:n))*h;
end
h2=(b-a)/100;
xc=a+(0:100)*h2;
fc=feval('function_f',xc);
plot(xc,fc,'r');
hold on; title('Trapezoidal rule');
xlabel('x'); ylabel('y'); grid on;
plot(x,f,'m');
plot(x,zeros(size(x)),'c')

for i=1:n;
    plot([x(i),x(i)],[0,f(i)],'g');
end
```

The function "function_f" is specified by the program "function_f.m".

For example, we can define "function_f" using the following subprogram, titled "function_f.m":

```
function f=f(x);
f=x+2*log(0.0000025235*x);
```

By executing the program trapez_g('function_f',1,50,60), where the bounds of integration have been defined as 1 and 50 and the number of intervals is 60, we obtain the integral of f over $[1, 50]$ with 60 steps:

```
>>trapez_g('function_f',1,50,60)
ans =
      279.3889
```

5.8.3. *Simpson's rule*

Suppose that we wish to evaluate the integral of $f(x) = \sqrt{1 + \exp x}$ over the interval $[0, 2]$ using Simpson's method for $n = 2$, $n = 4$, $n = 8$ and $n = 16$.

We will use the following script, titled "msimp.m":

```
clc; clear all;
Iexact=4.006994;
a=0;b=2;

fprintf('\n Simpson''s rule \n');

fprintf('\n n\t \t I\t Relat. error (perc.) \n');
fprintf('-------------------------------------------\n');
n=1;

for k=1:4
    n=2*n;
    h=(b-a)/n;
    i=1:n+1;
    x=a+(i-1)*h;
    f=sqrt(1+exp(x));
    I=h/3*(f(1)+4*sum(f(2:2:n))+f(n+1));
    if n>2
        I=I+h/3*2*sum(f(3:2:n));
    end
    errorI=abs(Iexact-I)/Iexact;
    fprintf('%d\t %10.5f\t %10.8f\n',n,I,errorI);
end
```

This algorithm produces the following results:

```
Simpson's rule

n            I              Relat. error (perc.)
--------------------------------------------------
2          4.00791          0.00022935

4          4.00705          0.00001521

8          4.00700          0.00000101

16         4.00699          0.00000012
```

It can be shown that, for $n > 16$, the relative error is very close to zero (of the order of 10^{-6}).

Let $f(x) = \dfrac{R^2 - x^2}{b^2\sqrt{g(x+R)}}$, where $n = 8$, $b = 0.1$, and $g = 9.81$. Define:

$$I = \int_{-0.9R}^{R} f(x)dx = \int_{-0.9R}^{R} \frac{(R^2 - x^2)}{b^2\sqrt{g(x+R)}}dx.$$

The following *Matlab* script computes the value of I using Simpson's rule. The script is saved as "simp.m":

```
clear all;
R=5; g=9.81; b=0.1;
x1=-0.90*R;
x2=R; h=(x2-x1)/100;
x=x1:h:x2;
f=(R^2-x.^2)./(b^2*sqrt(2*g*(x+R)));
I=simp_v(f,h)
```

The subprogram "simp_v(f,h)" is implemented as follows:

```
function I=simp_v(f,h);
n=length(f)-1; % Number of intervals

if n==1
    fprintf('Data only has one interval'\n');
    return;
end

if n==2
    I=h/3*(f(1)+4*f(2)+f(3));
```

```
    return; % Return and continue the calculation
end

if n==3
    I=(3/8)*h*(f(1)+3*f(2)+3*f(3)+f(4));
    return;
end

I=0;
if 2*floor(n/2)~=n ; % The number of intervals must be even
    I=3*h/8*(f(n-2)+3*f(n-1)+3*f(n)+f(n+1));
    m=n-3;
else
    m=n;
end

I=I+h/3*(f(1)+4*sum(f(2:2:m))+f(m+1));

if m>2
    I=I+(h/3)*2*sum(f(3:2:m))
end
```

Running the program "simp.m" allows us to find the value of the integral:

```
>>simp
I =
    1.8522e+003
```

PART 3

Solving Linear Systems

6

Matrix Norm and Conditioning

6.1. Introduction

We can define norms on matrices in the same way as for vectors. The second part of this book discusses matrix norms and some of their applications.

We will then consider one of the most essential problems encountered when running computer programs involving matrix analysis. As computers have finite memory, the results of their numerical computations are always slightly inaccurate. As these errors accumulate over large numbers of computations, or if they are combined with other errors introduced by approximation techniques, we need some way of judging the reliability of the results. The condition number of a matrix, which we shall define in section 6.3, is one example of a tool that we can use to estimate the reliability of numerical results.

6.2. Matrix norm

Norms can be defined on matrices in the same way as on vectors in $\mathbb{R}^n$.

DEFINITION.– *Suppose that A, B* $\in \mathcal{M}_{n\times m}$*, which denotes the set of* $n \times m$ *real matrices. The mapping* $\| \, . \, \|: \mathcal{M}_{n\times m} \to \mathbb{R}^+$ *is said to be a matrix norm if it satisfies the following properties:*

1) $\| A \| \geq 0$ *and* $\| A \| = 0 \Leftrightarrow A = 0$*;*

2) $\| \beta A \| = | \beta | \| A \|$ *for every scalar* β*;*

3) $\| A + B \| \leq \| A \| + \| B \|$*;*

4) $\| AB \| \leq \| A \| \| B \|$.

REMARK.– $\|A\| = \max_{1\le i,j\le 2} |a_{ij}|$ is a norm on $\mathcal{M}_{2\times 2}$ but is not a matrix norm. To verify this, consider the matrix:

$$A = \begin{pmatrix} 1 & 1 \\ 1 & 1 \end{pmatrix}.$$

This matrix satisfies $\|A^2\| = 2$, but $\|A\|^2 = 1$, so $\|A^2\| > \|A\|^2$.

Below, we give a few examples of matrix norms. The most common family of norms, again known as the p-norms, are derived from the vector norms defined earlier.

DEFINITION.– *Let A be an $n \times m$ matrix. Given any vector norm, the subordinate matrix norm is the matrix norm defined by:*

$$\| A \| = \max_{x\in\mathbb{R}^n,\ x\neq 0} \frac{\| Ax \|}{\| x \|} = \max_{\|x\|=1} \| Ax \| .$$

This relation means that the norm of a matrix is equal to the maximum value that can be obtained by multiplying A by every vector with norm 1. The norm of a given matrix A is not easy to compute from this equation, so the next few results are helpful in practice.

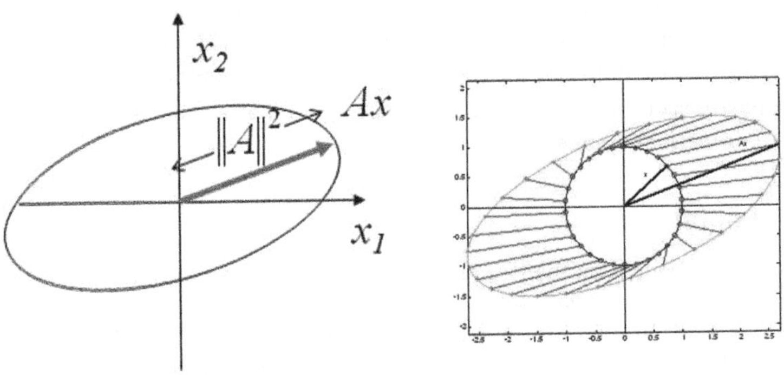

Figure 6.1. *Geometric illustration of the matrix norm* $\| A \| = \max_{\|x\|=1} \| Ax \|$

THEOREM.– *If $A = \{a_{ij}\} \in \mathcal{M}_{n\times m}$, the infinity norm satisfies the formula*

$$\| A \|_\infty = \max_{1\le i\le n} \sum_{j=1}^{m} | a_{ij} | .$$

A similar result holds for the 1-norm.

THEOREM.– *If* $A = \{a_{ij}\} \in M_{n\times m}$*, the* 1*-norm satisfies the formula*

$$\| A \|_1 = \max_{1\leq j\leq m} \sum_{i=1}^{n} | a_{ij} | .$$

REMARK.– These norms satisfy the property $\|A\|_1 = \|A^T\|_\infty$.

The formula for the 2-norm is more difficult to derive. We will state it without justification below. First, we need to define the notion of the spectral radius of a matrix.

DEFINITION.– *The spectral radius of a square matrix A, written as* $\rho(A)$*, is defined as follows:*

$$\rho(A) = \max_{i=1,\ldots,n} | \lambda_i |,$$

where the λ_i *are the eigenvalues of A. Note that the eigenvalues can be complex in general, in which case* $| \lambda_i |$ *denotes their modulus (which is a positive real number or zero).*

THEOREM.– *Suppose that A is an* $n \times m$ *matrix. Then:*

$$\| A \|_2 = \sqrt{\rho(A^T A)}.$$

LEMMA.– *If A is a symmetric* $n \times n$ *square matrix, then:*

$$\| A \|_2 = \rho(A).$$

REMARK.– In general, the spectral radius is not a norm: $A = \begin{pmatrix} 0 & 1 \\ 0 & 0 \end{pmatrix}$, $\| A \|_1 = 1 \neq 0$, but $\rho(A) = 0$.

EXAMPLE.–

1) Compute the values of the various norms of the matrix:

$$A = \begin{pmatrix} 2 & 2 \\ 1 & 3 \end{pmatrix} \quad A^T A = \begin{pmatrix} 5 & 7 \\ 7 & 13 \end{pmatrix}.$$

The 1-norm and ∞-norm have values $\|A\|_1 = 5$ and $\|A\|_\infty = 4$. The eigenvalues of $A^T A$ are $\lambda_1 = 0.93774$ and $\lambda_2 = 17.06226$, so $\|A\|_2 = \sqrt{17.06226} = 4.1306$.

2) Compute the 1, 2 and ∞-norms of the matrix:

$$A = \begin{pmatrix} 2 & 1 & 1 \\ 2 & 3 & 2 \\ 1 & 1 & 2 \end{pmatrix}.$$

The 1-norm and the infinity norm have values:

$$\| A \|_\infty = \max_{1\le i\le n} \sum_{j=1}^{n} | a_{ij} | = \max\{4,7,4\} = 7,$$

$$\| A \|_1 = \max_{1\le j\le n} \sum_{i=1}^{n} | a_{ij} | = \max\{5,5,5\} = 5.$$

As for the 2-norm, we first need to find the eigenvalues of $A^T A$. Therefore:

$$\begin{aligned} 0 &= \det(A^T A - \lambda I) \\ &= \begin{vmatrix} 9-\lambda & 9 & 8 \\ 9 & 11-\lambda & 9 \\ 8 & 9 & 9-\lambda \end{vmatrix} \\ &= 25 - 53\lambda + 29\lambda^2 - \lambda^3. \end{aligned}$$

Solving this equation yields the three eigenvalues of $A^T A$:

$$\{1, 14+3\sqrt{19}, 14-3\sqrt{19}\}.$$

Hence:

$$\begin{aligned} \| A \|_2 &= \sqrt{\rho(A^t A)} = \sqrt{\max\{1, 14+3\sqrt{19}, 14-3\sqrt{19}\}} \\ &= \sqrt{14+3\sqrt{19}} \approx 5.2035. \end{aligned}$$

Just as various other norms can be defined on vectors, there are other types of matrix norms, such as the Fröbenius norm, which can be defined as follows:

DEFINITION.– *Let $A \in M_{n\times m}(\mathbb{R})$. The Fröbenius norm is defined by:*

$$\| A \|_F = \sqrt{\sum_{i=1}^{n}\sum_{j=1}^{m} | a_{ij} |^2}.$$

THEOREM.–

1) $\| A \|_F = \sqrt{tr(A^T A)}$.

2) For every unitary matrix U, $\| UA \|_F = \| AU \|_F = \| A \|_F$.

REMARK.– The Fröbenius norm is easier to compute than the 2-norm. However, it requires a multiplicative coefficient, since $\| I \|_F = \sqrt{n}$, whereas $\| I \|_2 = 1$.

THEOREM.– $\forall A \in M_{n\times n}(\mathbb{R})$, $\| A \|_2 \leq \| A \|_F \leq \sqrt{n} \| A \|_2$.

LEMMA.– *For every* $A \in M_{n\times n}(\mathbb{R})$, *we have that* $\rho(A) \leq \| A \|_p$ *for* $p = 1, 2, \infty$, *and* $\rho(A) \leq \| A \|_F$.

REMARK.– If D is a diagonal matrix, then

$$\| D \|_1 = \| D \|_2 = \| D \|_\infty = \rho(D) \neq \| D \|_F .$$

6.3. Condition number of a matrix

Consider the following system:

$$\begin{pmatrix} 1 & 2 \\ 1.0001 & 2 \end{pmatrix} \begin{pmatrix} x_1 \\ x_2 \end{pmatrix} = \begin{pmatrix} 3 \\ 3 \end{pmatrix}.$$

This system admits two vectors x and y such that:

$$x = \begin{pmatrix} 1 \\ 1 \end{pmatrix} \quad \text{and} \quad y = \begin{pmatrix} 3 \\ 0 \end{pmatrix} \quad \begin{cases} r_x = Ax - b = (0, 0.0001)^T \\ r_y = Ay - b = (0, 0.0003)^T \end{cases}$$

Note that the vectors x and y are very different but give very close solutions.

This is because the matrix A is ill-conditioned. The condition number of a matrix is defined as follows:

DEFINITION.– *The condition number of a non-singular square matrix* A *for a given norm is defined as the value:*

$$K_p(A) = \|A\|_p \|A^{-1}\|_p.$$

REMARK.–

– The condition number describes how sensitive the solutions of the system are to perturbations in the data.

– The condition number does not depend on the choice of vector b.

– To calculate $K(A)$, we need to find A^{-1}.

– The following relation holds:

$$\| I \| = \| AA^{-1} \| \leq \| A \| \| A^{-1} \| \Rightarrow K(A) \geq 1.$$

A problem is said to be well-conditioned if the condition number is close to 1. For example, the condition number of the identity matrix is exactly 1, and the identity matrix is clearly extremely well-conditioned. The problem is said to be ill-conditioned if $K(A)$ is large (and ill-posed if $K(A)$ is infinite).

EXAMPLE.–

1) Suppose that we wish to calculate the condition number of the matrix:

$$A = \begin{pmatrix} 2 & 2 \\ 1 & 3 \end{pmatrix} \quad A^{-1} = \frac{1}{4}\begin{pmatrix} 3 & -2 \\ -1 & 2 \end{pmatrix}.$$

We computed the 2-norm $\|A\|_2 = 4.1306$ earlier. The 2-norm of the matrix A^{-1} is $\|A^{-1}\|_2 = 1.0327$. Therefore, $K_2(A) = 4.2656$.

To calculate $K_\infty(A)$, note that $\|A\|_\infty = 4$ and $\|A^{-1}\|_\infty = 1.25$, so $K_\infty(A) = 4 \times 1.25 = 5$.

2) The Hilbert matrix is one example of an ill-conditioned matrix:

$$H_n = \begin{pmatrix} 1 & \frac{1}{2} & \frac{1}{3} & \cdots & \frac{1}{n} \\ \frac{1}{2} & \frac{1}{3} & \frac{1}{4} & \cdots & \frac{1}{n+1} \\ \frac{1}{3} & \frac{1}{4} & \frac{1}{5} & \cdots & \frac{1}{n+2} \\ \cdots & \cdots & \cdots & \cdots & \cdots \\ \frac{1}{n} & \frac{1}{n+1} & \frac{1}{n+2} & \cdots & \frac{1}{2n-1} \end{pmatrix}.$$

The condition number of the Hilbert matrix is as follows:

n	2	3	4	5	6
$K_2(H_n)$	1.2×10^1	1.9×10^2	6.5×10^3	1.8×10^5	4.4×10^6

THEOREM.– *Suppose that A is an invertible square matrix:*

– *Perturbation of the matrix:*

$$(A + \delta A)(x + \delta x) = b \Rightarrow \frac{\| \delta x \|}{\| x + \delta x \|} \leq K(A)\frac{\| \delta A \|}{\| A \|}.$$

– *Perturbation of the right-hand side:*

$$A(x + \delta x) = b + \delta b \Rightarrow \frac{\| \delta x \|}{\| x \|} \leq K(A)\frac{\| \delta b \|}{\| b \|}.$$

EXAMPLE.– Consider the system $Ax = b$, where:

$$A = \begin{pmatrix} 8 & 10 & 10 & 10 \\ 1 & 5 & 6 & 1 \\ 7 & 10 & 4 & 7 \\ 7 & 8 & 1 & 7 \end{pmatrix}, \quad b = \begin{pmatrix} 38 \\ 13 \\ 28 \\ 23 \end{pmatrix}.$$

This system has the solution $x = \begin{pmatrix} 1 \\ 1 \\ 1 \\ 1 \end{pmatrix}$.

Now, consider the following perturbation of the vector b:

$$b + \delta b = \begin{pmatrix} 38.1 \\ 12.9 \\ 28.1 \\ 22.9 \end{pmatrix} \Rightarrow x + \delta x = \begin{pmatrix} -34.5 \\ 11.0 \\ -5.6 \\ 26.0 \end{pmatrix}.$$

Then:

$$\frac{\| \delta b \|}{\| b \|} = 3.7 \times 10^{-3} \qquad \frac{\| \delta x \|}{\| x \|} = 22.6.$$

The eigenvalues of the matrix A are $\lambda_1 = 24.1$ and $\lambda_4 = -0.007$. Therefore, the condition number of A is $K(A) = 7363$, so the system is ill-conditioned, which explains why the result differs so much from the true solution.

THEOREM.–

1) $K(A) = K(A^{-1})$.

2) $K_2(A) = \sqrt{\dfrac{\rho(A^T A)}{\epsilon(A^T A)}}$, *where* $\epsilon(B) = \min\limits_{\lambda \in S_p(B)} |\lambda|$.

3) If A is Hermitian, then $K_2(A) = \dfrac{\rho(A)}{\epsilon(A)}$.

4) If A is symmetric positive-definite, then $K_2(A) = \frac{\lambda_{max}}{\lambda_{min}} = \rho(A)\rho(A^{-1})$.

5) $\forall \alpha \in \mathbb{R}^*$, $K(\alpha A) = K(A)$.

6) If U is unitary, then $K_2(U) = 1$.

7) If U is unitary, then $K_2(UA) = K_2(A)$.

REMARK.– These properties allow us to make the following remarks:

– unitary matrices are among the matrices with the best conditioning;

– we cannot improve the conditioning by performing translations or dilations;

– we cannot improve the conditioning by multiplying with a unitary matrix.

EXAMPLE.– In certain special cases, the spectral condition number can be calculated from the eigenvalues. For example, consider the tridiagonal $N \times N$ matrix A defined by:

$$A = \begin{pmatrix} 2 & -1 & 0 & 0 & 0 \\ -1 & 2 & -1 & 0 & 0 \\ 0 & \ddots & \ddots & \ddots & 0 \\ 0 & 0 & -1 & 2 & -1 \\ 0 & 0 & 0 & -1 & 2 \end{pmatrix}.$$

Setting $h = \pi/(N+1)$ and using the trigonometric formula:

$$\sin(\alpha - \beta) + \sin(\alpha + \beta) = 2 \sin \alpha \cos \beta,$$

we can verify that the eigenvalues of A are:

$$\lambda_k = 2 - 2 \cos kh, \quad 1 \leq k \leq N,$$

and the corresponding eigenvectors are:

$$\vec{v}_k = (\sin kh, \sin 2kh, \ldots, \sin Nkh)^T, \quad 1 \leq k \leq N.$$

As the matrix A is symmetric and has positive eigenvalues, it is positive definitive and so its spectral condition number $K(A)$ is:

$$K(A) = \frac{1 - \cos Nh}{1 - \cos h} = \frac{1 + \cos h}{1 - \cos h}.$$

When N is large, $\cos h$ behaves like $1 - h^2/2$, and so $K(A) \approx 4/h^2 = O(N^2)$. Therefore, as the order N of the matrix A increases, the conditioning of the linear system $Ax = b$ deteriorates.

6.3.1. *Approximation of* $K(A)$

When performing matrix analyses, we sometimes need to compute the condition number of a matrix. However, this requires us to compute the norm of the inverse matrix A^{-1}, which is expensive in general, of order $O(n^3)$. Instead, it might be preferable to estimate $K(A)$. There exist algorithms with a complexity of order $O(n^2)$ to calculate the value of $K(A)$.

6.4. Preconditioning

There is no single method for preconditioning linear systems that always works, but there are several more or less empirical methods that we can use. For example, given a matrix with a strongly dominant diagonal (i.e. $\forall i = 1, \ldots, n$, $| a_{ii} | \gg \sum_{j=1}^{n} | a_{ij} |$), we can multiply by the diagonal matrix whose diagonal elements are the inverses of the diagonal elements of A.

Every preconditioning method follows the same principle; the idea is to find a matrix that is as close as possible to A^{-1}, and multiply the system by this matrix. Thus, we can find a suitable "preconditioner" that is adapted to the model we are working with.

6.5. Using *Matlab*

6.5.1. *Matrices and vectors*

Square matrices have the same number of rows and columns ($n = m$). A row matrix is a matrix of order 1, said to be a row vector. A column matrix is a matrix of order 1, said to be a column vector. A matrix of order 0 is a scalar.

The zero matrix is the matrix whose coefficients are all zero. By writing "O" for this matrix, we can construct O in *Matlab* as follows:

```
>>A=zeros(n,m);
```

The parameters n and m denote the number of rows and the number of columns respectively.

The function eye in *Matlab* returns the identity matrix.

EXAMPLE.–

$$I = \begin{pmatrix} 1\,0\,0\,0\,0 \\ 0\,1\,0\,0\,0 \\ 0\,0\,1\,0\,0 \\ 0\,0\,0\,1\,0 \\ 0\,0\,0\,0\,1 \end{pmatrix}$$

This is a square matrix of order 5. We can construct this matrix in *Matlab* by running the command:

```
>>I=eye(5)
I =
    1 0 0 0 0
    0 1 0 0 0
    0 0 1 0 0
    0 0 0 1 0
    0 0 0 0 1
```

Consider the matrix:

$$A = \begin{pmatrix} 1 & 3 & 5 \\ 2 & -1 & 0 \\ 5 & 4 & 3 \end{pmatrix}.$$

There are two ways to write this matrix A in *Matlab*:

```
>>A=[1,3,5;2,-1,0;5,4,3];
```

and:

```
>>A=[1 3 5;2 -1 0;5 4 3];
```

Rows are separated by semi-colons, and columns are separated by space or commas. For example:

```
>>A=[1 3 5;2 -1 0;5 4 3]
A =
     1     3     5
     2    -1     0
     5     4     3
```

We can write the matrix transpose A^T of A in *Matlab* as follows:

```
>>B=A'  % the apostrophe denotes the transpose
B =
     1     2     5
     3    -1     4
     5     0     3
```

Matrix multiplication is written as $*$. The determinant of a square matrix A can, for example, be obtained by executing:

```
>>det(A)
ans =
    44
```

Let A be a non-zero matrix. The inverse matrix A^{-1} of A (if it exists) satisfies $A * A^{-1} = Id$. In *Matlab*, the inverse matrix is returned by $A^(- 1) = inv(A) = pinv(A)$.

The inverse matrix of A is:

```
>>inv(A)
ans =
   -0.0682    0.2500    0.1136
   -0.1364   -0.5000    0.2273
    0.2955    0.2500   -0.1591
```

Alternatively:

```
>>pinv(A);
```

6.5.2. *Condition number of a matrix*

Consider the following system of n equations and m unknowns:

$$Ax = y,$$

where $A = (a_{ij})_{1 \leq i \leq n, 1 \leq j \leq m}$, $x = (x_i)_{1 \leq i \leq m}$ and $y = (y_j)_{1 \leq j \leq m}$.

– If $n > m$, the system is said to be overdetermined.

– If $n = m$, the system is said to be a Cramer system and has a unique solution (provided that $det(A) \neq 0$).

– If $n < m$, the system is said to be underdetermined.

Solving this system numerically (in the case $n = m$) allows us to find the unknowns x_i: $x = A/y$.

For any given column vector y (e.g. $y^T = (6, 0, -1)$), we can find a solution with *Matlab* as follows:

```
>> A = [1 3 5;2 -1 0;5 4 3];
>> y = [6 0 -1]';
>>x=A^(-1)*y
x =
    -0.5227
    -1.0455
     1.9318
```

To see the types of problems that can arise from poorly-conditioned square matrices, consider the following example.

EXAMPLE.– The effect of rounding errors is illustrated by the following example system of two equations:

$$\begin{cases} 0.12065x_1 + 0.98775x_2 = 2.01045 \\ 0.12032x_1 + 0.98775x_2 = 2.00555 \end{cases}$$

This system has the solutions $x_1 = 14.7403$ and $x_2 = 0.23942$.

To find the error introduced into the solutions by rounding the coefficients, suppose that we slightly increase the coefficient 2.01045 by 0.001 in the first equation.

The new system is:

$$\begin{cases} 0.12065x_1 + 0.98775x_2 = 2.01145 \\ 0.12032x_1 + 0.98775x_2 = 2.00555 \end{cases}$$

This will show the error introduced by rounding.

The matrix A of this new system is:

$$A = \begin{pmatrix} 0.12065 & 0.98775 \\ 0.12032 & 0.98775 \end{pmatrix}.$$

The norm of A is:

$$\|A\| = \left(\sum_{ij} |a_{ij}^2| \right)^{1/2} = 1.97993839.$$

In *Matlab*, the condition number of the matrix A is returned by the command *cond(A)*. This number always satisfies the condition $\text{cond}(A) \geq 1$. Ill-conditioned matrices can have large condition numbers, but this does not directly estimate the error in the solution.

```
>>A = [0.12065 0.98775; 0.12032 0.98775]

A =
    0.1206    0.9878
    0.1203    0.9878
>>cond(A)

ans =
  6.0754e+003
```

Consider the Hilbert matrix defined earlier, which is an example of an ill-conditioned matrix. The coefficients of this matrix satisfy:

$$A = (a_{ij}), \text{ where } a_{ij} = \frac{1}{i+j-1}.$$

The following script calculates the condition number (cond(A)) of a Hilbert matrix A, and also computes the value det(A)det(A^{-1}):

```
clear all;
for n=5:14
 for i=1:n
  for j=1:n
    A(i,j)=1/(i+j-1);
  end
 end
K=cond(A);
d=det(A)*det(A^(-1));
fprintf('n=%3.0f\t cond(a)=%e\t det*det=%e\n',n,K,d);
end
```

We can use this script to compute the condition numbers of the Hilbert matrices with orders 5 to 14:

```
n=  5     cond(a)=4.766073e+005    det*det=1.000000e+000

n=  6 cond(a)=1.495106e+007    det*det=1.000000e+000

n=  7 cond(a)=4.753674e+008    det*det=1.000000e+000

n=  8 cond(a)=1.525758e+010    det*det=1.000000e+000

n=  9 cond(a)=4.931542e+011    det*det=1.000002e+000

n= 10 cond(a)=1.602468e+013    det*det=9.999290e-001

n= 11 cond(a)=5.226017e+014    det*det=9.985187e-001

Warning: Matrix is close to singular or badly scaled.
         Results may be inaccurate. RCOND = 2.692153e-017.
n= 12    cond(a)=1.699490e+016    det*det=9.595235e-001

Warning: Matrix is close to singular or badly scaled.
         Results may be inaccurate. RCOND = 2.739612e-018.
n= 13    cond(a)=3.459085e+017    det*det=1.092976e+000

Warning: Matrix is close to singular or badly scaled.
         Results may be inaccurate. RCOND = 2.448199e-019.
n= 14    cond(a)=4.692946e+017    det*det=-1.457475e+000
```

The following message is displayed for $n = 11$ to $n = 14$:

```
Warning: Matrix is close to singular or badly scaled.
         Results may be inaccurate. RCOND = 3.659249e-017.
```

This means that, from $n = 11$ to $n = 14$, the calculation finds that $\det(A) \times \det(A^{-1})$ is not equal to 1, indicating that rounding errors are starting to take effect.

7

Direct Methods

7.1. Introduction

We often need to solve linear systems when studying problems derived from physics. For example, to solve more complex numerical analysis problems (e.g. partial differential equations), one common approach is to reduce the problem to a linear system of n equations and n unknowns of the form:

$$(S) \qquad \begin{cases} a_{11}x_1 + a_{12}x_2 + \cdots + a_{1n}x_n &= b_1 \\ a_{21}x_1 + a_{22}x_2 + \cdots + a_{2n}x_n &= b_2 \\ \qquad\qquad . &= . \\ \qquad\qquad . &= . \\ \qquad\qquad . &= . \\ a_{n1}x_1 + a_{n2}x_2 + \cdots + a_{nn}x_n &= b_n \end{cases}$$

where a_{ij} $(1 \leq i \leq n\,;1 \leq j \leq n)$ and $b_i(1 \leq i \leq n)$ are given real numbers and $x_i(1 \leq i \leq n)$ are the unknowns.

In matrix notation, the linear system (S) may be written more simply as $Ax = b$, where A is the square matrix formed by the elements a_{ij}, and x and b are the column matrices formed by the elements x_i and b_i respectively [LAS 94a].

7.2. Method of determinants or Cramer's method

THEOREM.– *If* $\det A \neq 0$, *then the linear system* $Ax = b$ *has the unique solution*

$$x_i = \frac{\det \Delta_i}{\det A}, \quad i = 1, \ldots, n, \qquad [7.1]$$

where Δ_i *is the matrix obtained by replacing the i-th column of A by b.*

This method is usually only useful in theoretical contexts. In practice, it can only be used in small dimensions ($n \leq 4$), since it requires an extremely high number of operations (approximately $n(n+1)!$ operations).

7.2.1. *Matrix inversion by Cramer's method*

By Cramer's method, a matrix A of rank n is only invertible if its determinant Δ is non-zero. If so, multiplying A by its inverse matrix A^{-1} yields the identity matrix I:

$$A^{-1}A = AA^{-1} = I, \quad \text{where} \quad A\,I = A.$$

By applying Cramer's method to the matrix A, we can find A^{-1}.

EXAMPLE.– Consider the matrix

$$A = \begin{pmatrix} 1\;3\;3 \\ 2\;2\;0 \\ 3\;2\;6 \end{pmatrix}.$$

Cramer's method allows us to find: $A^{-1} = \frac{1}{5}\begin{pmatrix} -2 & 2 & 1 \\ 2 & \frac{1}{2} & -1 \\ \frac{1}{3} & -\frac{7}{6} & \frac{2}{3} \end{pmatrix}$, which satisfies:

$$A^{-1}A = AA^{-1} = I.$$

7.3. Systems with upper triangular matrices

Suppose that A is upper triangular ($a_{ij} = 0$ for $i > j$).

THEOREM.– *If $a_{ii} \neq 0$, $\forall i$, then the linear triangular system A has the unique solution:*

$$x_n = \frac{b_n}{a_{nn}},$$

$$x_i = \left(b_i - \sum_{j=i+1}^{n} a_{ij}x_j\right) / a_{ii}, \quad i = n-1, \cdots, 1. \qquad [7.2]$$

If the matrix A is lower triangular (i.e. $a_{ij} = 0$ for $i < j$), then the system has the following solution:

$$x_1 = \frac{b_1}{a_{11}},$$

$$x_i = \left(b_i - \sum_{j=1}^{i-1} a_{ij}x_j\right) / a_{ii}, \quad i = 2, \ldots, n. \qquad [7.3]$$

This method requires approximately n^2 operations.

7.4. Gaussian method

The Gaussian method transforms the linear system $Ax = b$ into an equivalent linear system $A'x = b'$, where A' is an upper triangular matrix, without changing the solution.

This is done by applying the following elementary transformations, which return linear systems that are equivalent to the original system:

– permutation of two rows;

– multiplication of a row by a non-zero constant;

– addition of one row multiplied by a non-zero constant to another row.

We define the matrix $\mathcal{A} = (a_{ij})_{1\leq i\leq n, 1\leq j\leq n+1}$ of the system (S) by $a_{i,n+1} = b_i$.

An equivalent triangular system can be found by applying $(n-1)$ steps of the Gaussian algorithm to the rows of the matrix $\mathcal{A}$.

The k-th step is defined as follows:

CASE 1.– If $a_{kk} \neq 0$, we leave rows 1 to k unchanged and construct a new matrix from $\mathcal{A}$ as follows:

$$a_{ij} \leftarrow a_{ij} - \frac{a_{ik}}{a_{kk}} a_{kj}, \quad i = k+1, \ldots, n\,; \quad j = k, \ldots, n+1. \qquad [7.4]$$

CASE 2.– If $a_{kk} = 0$, we exchange the k-th row with some other row l $(k+1 \leq l \leq n)$ such that $a_{lk} \neq 0$. The rows 1 to $k-1$ are left unchanged, and we transform the rest of the matrix A as before, namely:

$$a_{kj} \longleftrightarrow a_{lj}, \quad j = k, \cdots, n+1 \qquad [7.5]$$

and:

$$a_{ij} \leftarrow a_{ij} - \frac{a_{ik}}{a_{kk}} a_{kj}, \quad i = k+1, \cdots, n\,; \quad j = k, \cdots, n+1. \qquad [7.6]$$

CASE 3.– If $a_{kk} = 0$, and $a_{lk} = 0\ \forall k \leq l \leq n$, the algorithm stops (which means that the system is singular).

After the k-th step, the matrix $\mathcal{A}$ is of the form:

$$\mathcal{A}^{(k)} = \begin{pmatrix} a_{11} & a_{12} & \cdots & \cdot & & \cdot & \cdots & a_{1n} & a_{1,n+1} \\ 0 & a_{22} & \cdots & \cdot & & \cdot & \cdots & a_{2n} & a_{2,n+1} \\ \vdots & \ddots & \ddots & & & & & \vdots & \vdots \\ \vdots & & & \ddots & a_{kk} & a_{k,k+1} & \cdots & a_{kn} & a_{k,n+1} \\ \vdots & & & & 0 & a_{k+1,k+1} & & & \vdots \\ \vdots & & & & \vdots & \vdots & & & \vdots \\ 0 & \cdots & \cdots & & 0 & a_{n,k+1} & \cdots & a_{nn} & a_{n,n+1} \end{pmatrix},$$

where $a_{ii} \neq 0, \forall\, 1 \leq i \leq k$.

Finally, after $(n-1)$ iterations, the matrix $\mathcal{A}$ is of the form:

$$\mathcal{A}^{(n-1)} = \begin{pmatrix} a_{11} & \cdots & \cdots & \cdots & a_{1n} & a_{1,n+1} \\ 0 & \ddots & & & \vdots & \vdots \\ \vdots & \ddots & \ddots & & \vdots & \vdots \\ \vdots & & \ddots & \ddots & \vdots & \vdots \\ 0 & 0 & \cdots & 0 & a_{n,n} & a_{n,n+1} \end{pmatrix},$$

and the corresponding linear system is upper triangular, with the same solution as the original system.

In the k-th step, the element a_{kk} is called the pivot. This pivot cannot be equal to zero. To improve the accuracy of the computations, we need to maximize the distance between the pivot and zero, i.e. we need to choose the pivot with the highest absolute value.

Thus, in the k-th step, we find the row l that satisfies $|a_{lk}| = \max\limits_{k \leq i \leq n} |a_{ik}|$.

If $|a_{lk}| = 0$, the process is terminated (singular system). Otherwise, we exchange the k-th row with the l-th row, and transform the matrix as described above. This gives us the following theorem:

THEOREM.– *The Gaussian method can be applied to the system $Ax = b$ if and only if all of the pivots $a_{kk}^{(k)}$ are non-zero. If so, the system is equivalent to $A'x = b'$, where A' and b' are constructed by applying the Gaussian algorithm to A and b. Furthermore, $det(A) = \prod_{k=1}^{n} a_{kk}^{(k)}$.*

EXAMPLE.–

$$(S) \qquad \begin{cases} 4x_1 + x_2 - 2x_3 &= 1 & L_1 \\ -x_1 - 3x_2 + 4x_3 &= 2 & L_2 \\ x_1 + x_2 + 3x_3 &= 0 & L_3 \end{cases}$$

Step 1.

L_1 unchanged,

$L_2 \leftarrow L_2 + \frac{L_1}{4}$,

$L_3 \leftarrow L_3 - \frac{L_1}{4}$.

This gives:

$$(S) \qquad \begin{cases} 4x_1 + x_2 - 2x_3 = 1 \\ -11x_2 + 14x_3 = 9 \\ 3x_2 + 14x_3 = -1 \end{cases}$$

Step 2.

L_1, L_2 unchanged,

$L_3 \leftarrow L_3 + \frac{3}{11} L_1$.

This gives:

$$(S) \qquad \begin{cases} 4x_1 + x_2 - 2x_3 = 1 \\ -11x_2 + 14x_3 = 9 \\ 196x_3 = 16 \end{cases}$$

which gives the solution:

$$x_3 = \frac{4}{49}, \qquad x_2 = \frac{-35}{49}, \qquad x_1 = \frac{23}{49}.$$

EXAMPLE.–

$$(S) \qquad \begin{cases} x_2 + 4x_3 = 1 \\ 3x_1 + 3x_2 + x_3 = 0 \\ 4x_1 + 4x_2 - x_3 = 2 \end{cases}$$

We define:

$$\mathcal{A} = \begin{pmatrix} 0 & 1 & 4 & 1 \\ 3 & 3 & 1 & 0 \\ 4 & 4 & -1 & 2 \end{pmatrix}.$$

Step 1.

– Note that $a_{11} = 0$, so we need to exchange the first row of A with the second or third row.

– $|a_{31}| = \max_{1\leq i\leq 3} |a_{i1}|$, so we exchange the first row with the third row. This transforms A into:

$$\mathcal{A} = \begin{pmatrix} 4 & 4 & -1 & 2 \\ 3 & 3 & 1 & 0 \\ 0 & 1 & 4 & 1 \end{pmatrix}.$$

– Finally, $a_{21} \neq 0$ and $a_{31} = 0$, so we only transform the second row. This gives:

$$\mathcal{A} = \begin{pmatrix} 4 & 4 & -1 & 2 \\ 0 & 0 & 7 & -6 \\ 0 & 1 & 4 & 1 \end{pmatrix}.$$

Step 2.

– Again, $a_{22} = 0$, so we need to exchange the second row with the third row. This gives:

$$\mathcal{A} = \begin{pmatrix} 4 & 4 & -1 & 2 \\ 0 & 1 & 4 & 1 \\ 0 & 0 & 7 & -6 \end{pmatrix}.$$

– Next, $a_{32} = 0$, so we do not need to change the third row.

This gives a Gaussian matrix corresponding to the linear system:

$$(S) \qquad \begin{cases} 4x_1 + 4x_2 - x_3 = 2 \\ x_2 + 4x_3 = 1 \\ 7x_3 = -6 \end{cases}$$

which has the solution:

$$x_3 = \frac{-6}{7}, \qquad x_2 = \frac{31}{7}, \qquad x_1 = \frac{-29}{7}. \qquad [7.7]$$

7.4.1. *Solving multiple systems in parallel*

Suppose that we need to solve the p distinct linear systems $Ax = b_i$, $i = 1, \ldots, p$. We can solve them in parallel by formally constructing the matrix $\mathcal{M}_{n,n+p}(\mathbb{R})$: $[A|b_1|\cdots|b_p]$ and applying Gaussian transformations to this matrix to obtain $[A'|b'_1|\cdots|b'_p]$, where A' is upper triangular.

However, this process requires p back-substitutions, and so the number of computations required is identical in practice. We can use the Gauss–Jordan method to be more efficient.

7.5. Gauss–Jordan method

THEOREM.– *Let A be an arbitrary square matrix. There exist matrices S and D such that $SA = D$ and D is a diagonal matrix of order n.*

7.5.1. *Underlying principle*

The Gauss–Jordan method transforms the system $Ax = b$ into the system $A'x = b'$, where $A' = I_n$. It is an extension of the Gaussian method. The idea is to continue performing elementary operations after putting the matrix in triangular form until the identity matrix has been obtained.

Suppose that we apply this method to the matrix $[A|b]$ that represents the system $Ax = b$. The Gaussian method transforms this matrix into $[A'|b']$, which represents the equivalent system $A'x = b'$. Continuing this transformation, the Gauss–Jordan method leads us to $[I_n|\bar{x}]$, where $\bar{x}$ is the solution of the system.

This diagonalization is executed in n steps, each consisting of normalization and reduction.

Step 1.

a) Normalization: This operation normalizes a_{11} to 1 by premultiplying the matrix $[A, b]$ by $\left(\dfrac{1}{a_{11}}\right)$. The first row of $[A, b]$ is now:

$$a_{1j}^{(1)} = \frac{a_{1j}}{a_{11}} \quad j = 1, \ldots, n+1.$$

b) Reduction: The off-diagonal coefficients of A are reduced to $a_{21}^{(1)} = a_{31}^{(1)} = \cdots = a_{n1}^{(1)} = 0$.

This is done by premultiplying the matrix $[A, b]$ by $(-a_{21})\ldots(-a_{n1})$. Let:

$$a_{ij}^{(1)} = a_{ij} - a_{i1}a_{1j}^{(1)} \quad i = 2, \ldots, n \;;\; j = 1, \ldots, n+1.$$

The system $[A, b]^{(1)}$ may be written as:

$$\begin{pmatrix} 1 & a_{12}^{(1)} & \cdots & a_{1n}^{(1)} \\ 0 & a_{22}^{(1)} & \cdots & a_{2n}^{(1)} \\ \vdots & \vdots & \ddots & \vdots \\ 0 & a_{n2}^{(1)} & \cdots & a_{nn}^{(1)} \end{pmatrix} \begin{pmatrix} x_1 \\ x_2 \\ \vdots \\ x_n \end{pmatrix} = \begin{pmatrix} a_{1n+1}^{(1)} \\ a_{2n+1}^{(1)} \\ \vdots \\ a_{nn+1}^{(1)} \end{pmatrix}.$$

k-th step

a) Normalization: The pivot term $a_{kk}^{(k-1)}$ is normalized to 1 by premultiplying $[A, b]^{(k-1)}$ by $\dfrac{1}{a_{kk}^{(k-1)}}$.

b) Reduction: The off-diagonal terms of the k-th column are canceled by premultiplying $[A, b]^{(k-1)}$ by $-a_{ik}^{(k-1)}$. This can be written more concisely as:

$$a_{ij}^{(k)} = a_{ij}^{(k-1)} - a_{ik}^{(k-1)} a_{kj}^{(k-1)} \quad i = 1, \ldots, n \, ; \, i \neq k \, ; \, j = k+1, \ldots, n+1.$$

EXAMPLE.– Solve the following system using the Gauss–Jordan method:

$$\begin{pmatrix} 1 & 3 & 3 \\ 2 & 2 & 0 \\ 3 & 2 & 6 \end{pmatrix} \begin{pmatrix} x_1 \\ x_2 \\ x_3 \end{pmatrix} = \begin{pmatrix} 0 \\ 2 \\ 11 \end{pmatrix}.$$

Let $[A, b] = \begin{pmatrix} 1 & 3 & 3 & 0 \\ 2 & 2 & 0 & 2 \\ 3 & 2 & 6 & 11 \end{pmatrix}$.

$$k = 1 \xrightarrow{\text{normalization}} \begin{pmatrix} 1 & 3 & 3 & 0 \\ 2 & 2 & 0 & 2 \\ 3 & 2 & 6 & 11 \end{pmatrix} \xrightarrow{\text{reduction}} \begin{pmatrix} 1 & 3 & 3 & 0 \\ 0 & -4 & -6 & 2 \\ 0 & -7 & -3 & 11 \end{pmatrix},$$

$$k = 2 \xrightarrow{\text{normalization}} \begin{pmatrix} 1 & 3 & 3 & 0 \\ 0 & 1 & \frac{3}{2} & \frac{-1}{2} \\ 0 & -7 & -3 & 11 \end{pmatrix} \xrightarrow{\text{reduction}} \begin{pmatrix} 1 & 0 & \frac{-3}{2} & \frac{3}{2} \\ 0 & 1 & \frac{3}{2} & \frac{-1}{2} \\ 0 & 0 & \frac{15}{2} & \frac{15}{2} \end{pmatrix},$$

$$k = 3 \xrightarrow{\text{normalization}} \begin{pmatrix} 1 & 0 & \frac{-3}{2} & \frac{3}{2} \\ 0 & 1 & \frac{3}{2} & \frac{-1}{2} \\ 0 & 0 & 1 & 1 \end{pmatrix} \xrightarrow{\text{reduction}} \begin{pmatrix} 1 & 0 & 0 & 3 \\ 0 & 1 & 0 & -2 \\ 0 & 0 & 1 & 1 \end{pmatrix}.$$

The solution is $x^T = (3, -2, 1)$.

REMARK.– This method allows us to solve p linear systems $Ax = b_i$ with the same matrix A in parallel.

To do so, we first construct the matrix $[A|b_1|\cdots|b_n]$. The Gauss–Jordan method then transforms this matrix into $[I_n|x_1|\cdots|x_n]$, where the x_i are the preimages of the b_i under the matrix A, and so are the solutions of the p distinct linear systems $Ax = b_i$.

7.5.2. *Computing the inverse of a matrix with the Gauss–Jordan algorithm*

The Gauss–Jordan algorithm allows us to transform the matrix $C = [A, b]$ into the matrix $D = [I_n, x]$, where x is the solution of the linear system $Ax = b$, or in other words $x = A^{-1}b$. Now, if we augment A by adjoining the identity matrix I_n, i.e. by setting $C = [A, I_n]$, then, after performing the Gauss–Jordan algorithm, we obtain:

$$D = A^{-1}C = A^{-1}[A, I_n] = \left[I_n, A^{-1}\right].$$

EXAMPLE.– Suppose that we wish to compute the inverse of $A = \begin{pmatrix} 2 & -1 & 4 \\ 1 & 0 & 2 \\ -1 & 2 & -3 \end{pmatrix}$.

We begin by forming the matrix:

$$\left(\begin{array}{ccc|ccc} 2 & -1 & 4 & 1 & 0 & 0 \\ 1 & 0 & 2 & 0 & 1 & 0 \\ -1 & 2 & -3 & 0 & 0 & 1 \end{array}\right).$$

Reduce the first column of A and apply the corresponding operations to I_3:

$$\left(\begin{array}{ccc|ccc} 1 & -0.5 & 2 & 0.5 & 0 & 0 \\ 0 & 0.5 & 0 & -0.5 & 1 & 0 \\ 0 & 1.5 & -1 & 0.5 & 0 & 1 \end{array}\right).$$

Reduce the second column of A:

$$\left(\begin{array}{ccc|ccc} 1 & 0 & 2 & 0 & 1 & 0 \\ 0 & 1 & 0 & -1 & 2 & 0 \\ 0 & 0 & -1 & 2 & -3 & 1 \end{array}\right).$$

Reduce the third column of A:

$$\left(\begin{array}{ccc|ccc} 1 & 0 & 0 & 4 & -5 & 2 \\ 0 & 1 & 0 & -1 & 2 & 0 \\ 0 & 0 & 1 & 2 & -3 & 1 \end{array}\right).$$

Hence:

$$A^{-1} = \begin{pmatrix} 4 & -5 & 2 \\ -1 & 2 & 0 \\ -2 & 3 & -1 \end{pmatrix}.$$

7.6. LU decomposition

LU decomposition expresses the matrix A as the product of a lower triangular matrix L with units along the diagonal and an upper triangular matrix U:

$$A = LU,$$

using the following formulas:

$$\begin{array}{ll} u_{1j} = a_{1j} & \text{if } j = 1, \ldots, n \\ l_{i1} = \frac{a_{i1}}{u_{11}} & \text{if } i = 1, \ldots, n \\ l_{ii} = 1 & \\ l_{ij} = 0 & \text{if } j = i+1, \ldots, n \\ u_{ij} = 0 & \text{if } j = 1, \ldots, i-1 \\ u_{ij} = a_{ij} - \sum_{k=1}^{i-1} l_{ik} u_{kj} & \\ l_{ij} = \dfrac{a_{ij} - \sum\limits_{k=1}^{j-1} l_{ik} u_{kj}}{u_{jj}} & \end{array}$$

The system $Ax = b$ can then be written as:

$$LUx = b,$$

i.e.:

$$\begin{cases} Ly = b \\ Ux = y \end{cases}$$

We can solve the first system to find the vector y, and then solve the second system to find the vector x. The matrices of these systems are triangular, so they are easy to solve.

The LU method can also be used to compute the determinant of A, which is equal to the product of the diagonal elements of the matrix U, since:

$$\det(A) = \det(L) \times \det(U),$$

and $\det(L) = 1$. Note that the determinant of any triangular matrix is equal to the product of its diagonal elements.

EXAMPLE.– Suppose that we wish to solve the system $Ax = b$ by LU decomposition, where:

$$A = \begin{pmatrix} 8 & -6 & 2 \\ -4 & 11 & -7 \\ 4 & -7 & 6 \end{pmatrix} \quad b = \begin{pmatrix} 28 \\ -40 \\ 33 \end{pmatrix}.$$

The LU decomposition of the matrix A is:

$$A = LU = \begin{pmatrix} 2 & 0 & 0 \\ -1 & 2 & 0 \\ 1 & -1 & 1 \end{pmatrix} \begin{pmatrix} 4 & -3 & 1 \\ 0 & 4 & -3 \\ 0 & 0 & 2 \end{pmatrix}.$$

First, we solve the system $Ly = b$ by substitution:

$$\begin{aligned} 2y_1 = 28 &\Rightarrow y_1 = \tfrac{28}{2} = 14, \\ -y_1 + 2y_2 = -40 &\Rightarrow y_2 = \tfrac{-40+y_1}{2} = \tfrac{-40+14}{2} = -13, \\ y_1 - y_2 + y_3 = 33 &\Rightarrow y_3 = 33 - y_1 + y_2 = 33 - 14 - 13 = 6. \end{aligned}$$

Next, we find the solution x by solving the system $Ux = y$ by substitution:

$$\begin{aligned} 2x_3 = y_3 &\Rightarrow x_3 = \tfrac{y_3}{2} = \tfrac{6}{2} = 3, \\ 4x_2 - 3x_3 = y_2 &\Rightarrow x_2 = \tfrac{y_2+3x_3}{4} = \tfrac{-13+3(3)}{4} = -1, \\ 4x_1 - 3x_2 + x_3 = y_1 &\Rightarrow x_1 = \tfrac{y_1+3x_2-x_3}{4} = \tfrac{14+3(-1)-3}{4} = 2. \end{aligned}$$

Therefore, the solution is $x = (2, -1, 3)^T$.

7.7. Thomas algorithm

The Thomas algorithm can be applied to matrices of the following form:

$$A = \begin{pmatrix} a_1 & c_1 & 0 & \cdots & 0 \\ b_2 & a_2 & c_2 & \ddots & \vdots \\ 0 & \ddots & \ddots & \ddots & 0 \\ \vdots & \ddots & b_{n-1} & a_{n-1} & c_{n-1} \\ 0 & \cdots & 0 & b_n & c_n \end{pmatrix}.$$

It can be shown that there exist two bidiagonal matrices L and U of the form:

$$L = \begin{pmatrix} 1 & 0 & \cdots & 0 \\ \beta_2 & 1 & \ddots & \vdots \\ 0 & \ddots & \ddots & 0 \\ 0 & \cdots & \beta_n & 1 \end{pmatrix}$$

$$U = \begin{pmatrix} \alpha_1 & \gamma_1 & \cdots & 0 \\ 0 & \alpha_2 & \ddots & \vdots \\ 0 & \ddots & \ddots & \gamma_{n-1} \\ 0 & \cdots & \cdots & \alpha_n \end{pmatrix}$$

such that $A = LU$.

The coefficients α_i, β_i and γ_i can easily be calculated using the following relations:

$$\alpha_1 = a_1, \; \beta_i = \frac{bi}{\alpha_{i-1}}, \; \alpha_i = a_i - \beta_i c_{i-1}, \; \gamma_i = c_i. \qquad [7.8]$$

Solving the system $Ax = f$ is equivalent to solving the two bidiagonal systems $Ly = f$ and $Ux = y$, which satisfy the following formulas:

$$\begin{cases} (Ly = f) : y_1 = f_1, \; y_i = f_i - \beta_i y_{i-1}, \; i = 2, \cdots, n, \\ (Ux = y) : x_n = \dfrac{y_n}{\alpha_n}, \; x_i = \dfrac{y_i - \gamma_i x_{i+1}}{\alpha_i}, \; i = n-1, \ldots, 1. \end{cases} \qquad [7.9]$$

7.8. Cholesky decomposition

The method of Cholesky decomposition (or factorization) is a direct method that is useful when A is symmetric positive-definite. Note that a matrix $A \in \mathcal{M}_n(\mathbb{R})$ with coefficients $(a_{ij})_{1 \le i,j \le n}$ is said to be symmetric if $A^T = A$, where A^T denotes the transpose of the matrix A, and is said to be positive-definite if $(Ax, x) > 0$ for all $x \in \mathbb{R}^n$ such that $x \neq 0$. If A is symmetric positive-definite, then it is necessarily invertible.

THEOREM.– *If A is a symmetric positive-definite matrix, then it may be factorized into $A = LL^T$.*

By developing the equation $A = LL^T$, we find $a_{ij} = \sum_{k=1}^{n} l_{ik} l_{jk}$, and $l_{ik} = 0$ if $i < k$. Therefore:

$$a_{ij} = \sum_{k=1}^{\min(i,j)} l_{ik} l_{kj}, \quad 1 \le i, \; j \le n.$$

The i-th row of the upper triangular part of A may be written as:

$$a_{ij} = \sum_{k=1}^{i} l_{ik} l_{jk} \quad j = i, \ldots, n.$$

Let:

$$a_{ij} = \sum_{k=1}^{i-1} l_{ik} l_{jk} + l_{ii} l_{ji} \quad j = i, \ldots, n.$$

This gives the following algorithm:

$$\left.\begin{array}{ll} l_{ii} = \left[a_{ii} - \displaystyle\sum_{k=1}^{i-1} (l_{ik})^2 \right]^{\frac{1}{2}} & \\ & \\ l_{ji} = \dfrac{a_{ij} - \displaystyle\sum_{k=1}^{i-1} l_{ik} l_{jk}}{l_{ii}} & j = i+1, \ldots, n \end{array}\right\} i = 1, \ldots, n$$

EXAMPLE.– Consider the matrix A defined by:

$$A = \begin{pmatrix} 9 & 1 \\ 1 & 4 \end{pmatrix}.$$

For $i = 1$, $l_{11} = \sqrt{a_{11}} = 2$, $l_{21} = \frac{a_{21}}{l_{11}} = \frac{1}{2}$.

For $i = 2$, $l_{22} = \sqrt{a_{22} - l_{11}^2} = \sqrt{9 - \dfrac{1}{4}} = \dfrac{\sqrt{35}}{2}$.

Therefore, $L = \begin{pmatrix} 2 & 0 \\ \dfrac{1}{2} & \dfrac{\sqrt{35}}{2} \end{pmatrix}$.

EXAMPLE.– Consider the $n \times n$ tridiagonal matrix A defined by:

$$A = \begin{pmatrix} 2 & -1 & & & \\ -1 & 2 & -1 & & \\ & \ddots & \ddots & \ddots & \\ & & -1 & 2 & -1 \\ & & & -1 & 2 \end{pmatrix}.$$

The Cholesky decomposition of this matrix is $A = LL^T$, where:

$$L = \begin{pmatrix} d_1 & & & \\ e_1 & d_2 & & \\ & \ddots & \ddots & \\ & & \ddots & d_{n-1} \\ & & & e_{n-1} \; d_n \end{pmatrix}.$$

Computing the product LL^T and setting it equal to A yields the following relations:

$$d_1^2 = 2,\ e_1 d_1 = -1,\ e_1^2 + d_2^2 = 2,\ e_2 d_2 = -1,$$
$$e_2^2 + d_3^2 = 2,\ e_3 d_3 = -1,\ ...,\ e_{n-1}^2 + d_n^2 = 2.$$

Thus, $d_1 = \sqrt{2}$, and, for $j = 1, 2, \ldots, n-1$:

$$e_j = -\frac{1}{d_j} \quad d_{j+1} = \sqrt{2 - e_j^2}.$$

The following algorithm computes the elements d_j ($j = 1, \ldots, n$) and e_j ($j = 1, \ldots, n-1$). Starting with $d_1 = \sqrt{2}$:

$$\begin{cases} \text{for } j = 1 : n \text{ do} \\ \quad e_j := -\dfrac{1}{d_j} \\ d_{j+1} := \sqrt{2 - e_j^2} \end{cases}$$

7.9. Using *Matlab*

7.9.1. *Matrix operations*

Consider the matrix A defined by:

$$A = \begin{pmatrix} 1\,2\,3 \\ 2\,3\,4 \\ 4\,2\,5 \end{pmatrix}.$$

There are several ways to construct this matrix in *Matlab*. For example:

```
>> A = [1 2 3; 2 3 4; 4 2 5]
A =
    1 2 3
    2 3 4
    4 2 5
```

Alternatively, we can enter each row separately, validating with the 'Enter' key between any two consecutive rows:

```
>> A = [ 1 2 3
2 3 4
4 2 5]

A =
    1 2 3
    2 3 4
    4 2 5
```

We can also separate the elements in each row by commas:

```
>> A = [ 1, 2, 3
2, 3, 4
4, 2, 5]
```

The first index labels the rows, and the second index labels the columns. For example:

```
>> A(2,1)

ans =
     2
```

The number of rows and columns of a matrix are returned by the *size* function. For example:

```
>> size(A)

ans =
     3 3
```

Consider a matrix B with the same number of rows as the matrix A. We can juxtapose the matrices A and B by writing:

```
>> [A B]
```

We can append the matrix B to the bottom of the matrix A by transposing it:

```
>> [A ; B']
```

We can also extract submatrices from a given matrix. For example, the following command extracts the first row of A:

```
>> A(1,:)

ans =
     1 2 3
```

In the above command, the number 1 in parentheses tells *Matlab* to include "the first row", and the colon after the comma tells *Matlab* to include "every column". We can similarly use the colon to include "every row":

```
>> A(:,1)

ans =
     1
     2
     4
```

In the next example, the colon is used twice, first to include "rows 1 to 2", and then to include "columns 1 to 2".

```
>> A(1:2, 1:2)

ans =
       1 2
       2 3
```

We can use the standard operators in the natural way with matrices: +, -, $*$ and ^ (exponentiation).

The following command generates a matrix with the same dimensions as the matrix A, and sets every coefficient equal to one:

```
>> [m, n] = size(A) ; ones(m, n)

ans =
      1 1 1
      1 1 1
      1 1 1
```

This can also be written more concisely as:

```
>> ones(size(A))
```

To find the rank of a matrix, we can call the *rank* function as follows:

```
>> rank(A)

ans =
       3
```

7.9.2. *Systems of linear equations*

7.9.2.1. *Cramer's method*

Consider the following system of linear equations:

$$\begin{cases} x_1 + 2x_2 + 3x_3 = 4 \\ 2x_1 + 3x_2 + 4x_3 = 5 \\ 4x_1 + 2x_2 + 5x_3 = 1 \end{cases}$$

We can use Cramer's method to solve this system. First, we define the matrix A and the vector b:

```
>> A = [1 2 3; 2 3 4; 4 2 5];
>> b= [ 4; 5; 1];
```

We can use *Matlab* to calculate the determinant of the matrix by calling the *det* function:

```
>> det(A)

ans =
       -5
```

Cramer's method computes the x_i $(i = 1, 2, 3)$ as fractions of determinants. Let $D1$ be the matrix whose determinant appears in the numerator of x_1. We can construct $D1$ in *Matlab* as follows:

```
>> D1 = A; D1(:,1) = b

D1 =
      4 2 3
      5 3 4
      1 2 5
```

Similarly, let $D2$ and $D3$ be the matrices whose determinants appear in the numerators of x_2 and x_3 respectively. We can construct $D2$ and $D3$ in the same way:

```
>> D2 = A; D2(:,2) = b

D2 =
      1 4 3
      2 5 4
      4 1 5

>> D3= A; D3(:,3) = b

D3 =
      1 2 4
      2 3 5
      4 2 1
```

Cramer's method therefore leads to the solution:

```
>> X = [det(D1); det(D2); det(D3)]/det(A)

X =
     -1.4000
      1.8000
      0.6000
```

To verify this solution, we can multiply A by X:

```
>> A*X

ans =
        4.0000
        5.0000
        1.0000
```

which does indeed recover the vector b.

7.9.2.2. *Solving systems of linear equations*

Suppose that we wish to solve the following system:

$$\begin{cases} 2x_1 - x_2 = 2 \\ x_1 + x_2 \;\; = 5 \end{cases}$$

This system has the solutions $x_1 = \dfrac{7}{3}, x_2 = \dfrac{8}{3}$.

There are two possible approaches with *Matlab*. The first method uses the inverse of the matrix (this approach performs poorly in general). We can do this as follows:

```
>> A = [2 -1 ; 1 1 ];
>> B = [2 ; 5];
>> X = inv(A)*B

X =
    2.3333
    2.6667
```

The function *rats* returns a rational approximation of this solution:

```
>> rats(X)

ans =
        7/3
        8/3
```

The second method applies Gaussian elimination using the backslash operator:

```
>> A = [2 -1 ; 1 1 ];
>> B = [2 ; 5];
>> X = A\B

X =
    2.3333
    2.6667
```

7.9.2.3. *Other functions*

The following functions are also available in *Matlab*:

– [L,U] = lu(A) returns the LU decomposition of $A = LU$. The matrix U is upper triangular and the matrix L is lower triangular;

– [M,U,P] = lu(A) returns the same matrix U as above, but the matrix M is lower triangular, and P is the permutation matrix such that $M = P * L$. Note that $P * A = M * U$;

– L = chol(A) returns the matrix L in the Cholesky factorization $A = LL^T$.

7.9.2.4. *Gaussian elimination*

The following script implements Gaussian elimination to solve a linear system $Ax = b$:

```
function [x,det] = gauss(A,b)
% Solve A*x = b by Gaussian elimination and compute det(A).
% USE: [x,det] = gauss(A,b)
if size(b,2) > 1; b = b'; end
% b must be a column vector
n = length(b);
% Elimination phase
for k = 1:n-1
  for i= k+1:n
    if A(i,k) ~= 0
       lambda = A(i,k)/A(k,k);
       A(i,k+1:n) = A(i,k+1:n) - lambda*A(k,k+1:n);
       b(i)= b(i) - lambda*b(k);
    end
  end
end if nargout == 2;
  det = prod(diag(A));
end
% Solution phase
for k = n:-1:1
    b(k) = (b(k) - A(k,k+1:n)*b(k+1:n))/A(k,k);
end
x = b;
```

We can use this script to solve the following linear system:

$$A = \begin{pmatrix} 1 & 0 & 1 & 1 \\ 1 & 2 & 3 & 4 \\ 1 & 3 & 6 & 10 \\ 1 & 4 & 10 & 20 \end{pmatrix} \quad \text{and} \quad b = \begin{pmatrix} 8 \\ 30 \\ 65 \\ 119 \end{pmatrix}.$$

This returns the result:

```
>> A =[1 0 1 1;1 2 3 4;1 3 6 10;1 4 10 20]
A =
     1     0     1     1
     1     2     3     4
     1     3     6    10
     1     4    10    20

>> b=[8 30 65 119]'
b =
     8
    30
    65
   119
>> [x,det] = gauss(A,b)
x =
     1
     2
     3
     4
det =
     7
```

To verify this result, we can compute Ax.

```
>> y= A*x
y =
     8
    30
    65
   119
```

The following *Matlab* script solves linear systems using Gaussian elimination with row pivoting:

```
function x = gaussPiv(A,b)
% Solve A*x = b by Gaussian elimination with row pivoting.
% USE: x = gaussPiv(A,b)
if size(b,2) > 1;
b = b';
end
n = length(b);
s = zeros(n,1);
%----------Find the pivot----------
for i = 1:n;
```

```
s(i) = max(abs(A(i,1:n)));
end
%---------Exchange rows if necessary----------
for k = 1:n-1
[Amax,p] = max(abs(A(k:n,k))./s(k:n));
p = p + k - 1;
if Amax < eps;
error('Matrix is singular');
end
if p ~= k

b = swapRows(b,k,p);
s = swapRows(s,k,p);
A = swapRows(A,k,p);
end
%--------------Elimination step---------------
for i = k+1:n
if A(i,k) ~= 0
lambda = A(i,k)/A(k,k);

A(i,k+1:n) = A(i,k+1:n) - lambda*A(k,k+1:n);

b(i) = b(i) - lambda*b(k);
end
end
end
%------------Substitution phase----------
for k = n:-1:1
b(k) = (b(k) - A(k,k+1:n)*b(k+1:n))/A(k,k);
end

x = b;
```

The function *swapRows* can be implemented as follows:

```
function v = swapRows(v,i,j)
% Exchange the i-th and j-th rows of the vector or matrix v.
% USE: v = swapRows(v,i,j)
temp = v(i,:);
v(i,:) = v(j,:);
v(j,:) = temp;
```

7.9.2.5. *Cholesky's method*

The following script implements Cholesky's method:

```
function L = cholesky(A)
% Compute L in the Choleksy factorization A = LL'.
% USE: L = cholesky(A)
n = size(A,1);
for j = 1:n

temp = A(j,j) - dot(A(j,1:j-1),A(j,1:j-1)); if temp < 0.0
error('Matrix is not positive-definite') end A(j,j) = sqrt(temp);

for i = j+1:n

A(i,j)=(A(i,j) - dot(A(i,1:j-1),A(j,1:j-1)))/A(j,j);
end
end

L = tril(A)
```

The commands *tril* and *dot* are built-in *Matlab* functions.

We can run this script to find the Cholesky factorization of the matrix:

$$A = \begin{pmatrix} 1 & 2 & 3 & 4 \\ 2 & 5 & 1 & 10 \\ 3 & 1 & 35 & 5 \\ 4 & 10 & 5 & 45 \end{pmatrix}.$$

This returns the following matrix L:

```
>> L = cholesky(A)
L =
    1    0    0    0
    2    1    0    0
    3   -5    1    0
    4    2    3    4
```

The next script computes the inverse of the matrix L in the Cholesky factorization:

```
function Linv = invert(L)
% Invert the lower triangular matrix L.
n = size(L,1);
for j = 1:n-1
L(j,j) = 1/L(j,j);
for i = j+1:n
```

```
L(i,j) = -dot(L(i,j:i-1),
L(j:i-1,j)/L(i,i));
end
end

L(n,n) = 1/L(n,n);
Linv = L;
```

Executing this script returns:

```
>> invert(L)

ans =
     1.0000         0         0         0
    -2.0000    1.0000         0         0
   -13.0000    5.0000    1.0000         0
     9.7500   -4.2500   -0.7500    0.2500
```

8

Iterative Methods

8.1. Introduction

The main idea of an iterative method is to construct a sequence of vectors $x^{(k)}$ from an initial vector $x^{(0)}$. This sequence is designed to converge to the vector x that solves the system $Ax = b$, where $A \in \mathcal{M}_n(K)$, $\det(A) \neq 0$ ($K = \mathbb{R}$ or $\mathbb{C}$) and:

$$x = \lim_{k \to +\infty} x^{(k)}. \tag{8.1}$$

After choosing a norm $\|.\|$ on K^n, this condition can be written as:

$$\lim_{k \to +\infty} \|x^{(k)} - x\| = 0. \tag{8.2}$$

One issue with the criterion [8.2] is that it includes the unknown x. However:

$$x^{(k)} - x = A^{-1}Ax^{(k)} - A^{-1}Ax = A^{-1}(Ax^{(k)} - b),$$

so:

$$\|x^{(k)} - x\| \leq \|A^{-1}\|\|x^{(k)} - x\|.$$

Similarly, it is clear that:

$$\|Ax^{(k)} - b\| \leq \|A\|\|x^{(k)} - x\|.$$

Therefore, equation [8.2] is equivalent to:

$$\lim_{k \to +\infty} \|Ax^{(k)} - b\| = 0. \tag{8.3}$$

This new criterion [8.3] depends solely on the properties of the system [ORT 70, LAS 94b].

One example of an iterative strategy is to consider the linear recurrence relation:

$$x^{(k+1)} = Bx^{(k)} + g, \tag{8.4}$$

where B is said to be the iteration matrix of the iterative method (which depends on A) and g is a vector (which depends on b) such that:

$$x = Bx + g. \qquad [8.5]$$

Defining the error at the k-th step as:

$$e^{(k)} = x - x^{(k)}, \qquad [8.6]$$

we obtain the recurrence relation $e^{(k)} = Be^{(k-1)}$. Therefore, $e^{(k)} = B^k e^{(0)}$; $k = 0, 1, \dots$. It is possible to show the following result:

THEOREM 8.1.– *The condition* $\lim_{k\to\infty} e^{(k)} = 0$ *holds for all* $e^{(0)}$ *(and thus for all* $x^{(0)}$*) if and only if* $\rho(B) < 1$*, where* $\rho(B)$ *is the spectral radius of the matrix* B*. Note that the spectral radius is defined by*

$$\rho(B) = \max |\lambda_i(B)|, \qquad [8.7]$$

where $\lambda_i(B)$ *are the eigenvalues of the matrix* B*.*

The convergence properties of the iterative process only depend on the matrix B. The choice of the vector g determines whether the limit of the iterative process is a solution of the original system. Clearly, this is the case if and only if the two systems $(I - B)x = g$ and $Ax = b$ have the same solution. The condition may therefore be stated as

$$(I - B)^{-1} g = A^{-1} b.$$

REMARK.– The advantage of iterative methods over direct methods is that they are easier to program and require less memory space. However, they often require longer computation times.

8.2. Classical iterative techniques

One general technique for constructing iterative methods is to find the so-called splitting of the matrix A into $A = M - N$, where M and N are chosen in such a way that M is non-singular. The matrix M is called the preconditioning matrix. More precisely, given $x^{(0)}$, we can compute $x^{(k)}$ for $k \geq 1$ by solving the system:

$$Mx^{(k+1)} = Nx^{(k)} + b\ ;\ k \geq 0. \qquad [8.8]$$

Clearly, the exact solution x satisfies $Mx = Nx + b$, and therefore also $Ax = b$. The system [8.8] can also be written in the form [8.4], where $B = M^{-1}N$ and $g = M^{-1}b$ [JED 01].

DEFINITION.– *For each vector* $y \in K^n$*, we define the residual vector of* y*, denoted by* $r(y)$*, as the vector* $r(y) = Ay - b$*. The components* $r_1,\dots, r_n$ *of* $r(y)$ *are also said to be residuals.*

The following recurrence relation is equivalent to [8.8]:

$$M(x^{(k+1)} - x^{(k)}) = r^{(k)},\ k \geq 0, \qquad [8.9]$$

where $r^{(k)} = b - Ax^{(k)}$ is the residual vector at the k-th iteration. The above relation can be generalized to:

$$M(x^{(k+1)} - x^{(k)}) = \alpha_k r^{(k)}, \qquad [8.10]$$

where the parameters α_k (which can be chosen to be distinct for each iteration k) are introduced to accelerate the convergence. This is called the Richardson method. Relations [8.8], [8.9] and [8.10] show that we need to solve a linear system with matrix M at each iteration; therefore, this matrix M must have a favorable form to ensure that the cost of solving the system remains reasonable. For example, M can be chosen to be diagonal or triangular.

8.2.1. *Jacobi method*

Note that if the diagonal elements of A are non-zero, the linear system $Ax = b$ is equivalent to:

$$x_i = \frac{1}{a_{ii}}\left(b_i - \sum_{j=1, j\neq i}^{n} a_{ij}x_j\right), \quad i = 1, \ldots, n. \qquad [8.11]$$

Given an initial value $x^{(0)}$, we can compute $x^{(k+1)}$ using:

$$x_i^{(k+1)} = \frac{1}{a_{ii}}\left(b_i - \sum_{\substack{j=1 \\ j\neq i}}^{n} a_{ij}x_j^{(k)}\right), \quad i = 1, \ldots, n. \qquad [8.12]$$

This allows us to split A into $A = D + E + F$, where D is the diagonal matrix of A, E is the lower triangular matrix with zeros along the diagonal and F is the upper triangular matrix with zeroes along the diagonal.

The iteration matrix of the Jacobi method is:

$$B_J = D^{-1}(E + F). \qquad [8.13]$$

The Jacobi algorithm requires the two vectors $x^{(k)}$ and $x^{(k+1)}$ to be stored in memory.

EXAMPLE.– Suppose that we wish to solve the following system using the Jacobi method:

$$\begin{cases} 10x_1 - x_2 + 2x_3 + \quad = 6 \\ -x_1 + 11x_2 - x_3 + 3x_4 = 25 \\ 2x_1 - x_2 + 10x_3 - x_4 = -11 \\ 3x_2 - x_3 + 8x_4 = 15 \end{cases} \qquad [8.14]$$

The exact solution of this system is $(1, 2, -1, 1)^T$. If we begin the Jacobi method with the initial vector $x^{(0)} = (0, 0, 0, 0)^T$, then $x^{(1)}$ can be computed as follows:

$$\begin{cases} x_1^{(1)} = \frac{1}{10}x_2^{(0)} - \frac{1}{5}x_3^{(0)} + \frac{3}{5} = 0.6 \\ x_2^{(1)} = \frac{1}{11}x_1^{(0)} + \frac{1}{11}x_3^{(0)} - \frac{3}{11}x_4^{(0)} + \frac{25}{11} = 2.2727 \\ x_3^{(1)} = -\frac{1}{5}x_1^{(0)} + \frac{1}{10}x_2^{(0)} + \frac{1}{10}x_4^{(0)} - \frac{11}{10} = -1.1 \\ x_4^{(1)} = -\frac{3}{8}x_2^{(0)} + \frac{1}{8}x_3^{(0)} + \frac{15}{8} = 1.875 \end{cases}$$

Table 8.1 displays each iteration up to $n = 10$, which satisfies the convergence criterion.

k	$x_1^{(k)}$	$x_2^{(k)}$	$x_3^{(k)}$	$x_4^{(k)}$
0	0.0000	0.0000	0.0000	0.0000
1	0.6000	2.2727	-1.1000	1.8750
2	1.0473	1.7159	-0.8052	0.8852
3	0.9326	2.0553	-1.0493	1.1309
4	1.0152	1.9537	-0.9681	0.9739
5	0.9890	2.0114	-1.0103	1.0214
6	1.0032	1.9922	-0.9945	0.9944
7	0.9981	2.0023	-1.0020	1.0036
8	1.0006	1.9987	-0.9990	0.9989
9	0.9997	2.0004	-1.0004	1.0006
10	1.0001	1.9998	-0.9998	0.9998

Table 8.1. *Results of each iteration of the Jacobi method*

The iterations are halted once the following criterion is met:

$$\frac{\|x^{(10)} - x^{(9)}\|_\infty}{\|x^{(10)}\|_\infty} = \frac{8 \times 10^{-4}}{1.9998} < 10^{-3},$$

which implies that $\|x^{(10)} - x\|_\infty \approx 0.0002$.

8.2.2. *Gauss–Seidel method*

The Gauss–Seidel method defines:

$$x_i^{(k+1)} = \frac{1}{a_{ii}}\left(b_i - \sum_{j=1}^{i-1} a_{ij}x_j^{(k+1)} - \sum_{j=i+1}^{n} a_{ij}x_j^{(k)}\right), \quad i = 1, \ldots, n. \quad [8.15]$$

This can be rewritten as:

$$x^{(k+1)} = (D-E)^{-1}(b + Fx^{(k)}). \quad [8.16]$$

The following splitting is used for A:

$$M = D - E, \ N = F, \quad [8.17]$$

and the corresponding iteration matrix is:

$$B_{GS} = (D-E)^{-1}F. \quad [8.18]$$

The Gauss–Seidel algorithm only requires one single vector to be stored in memory, as $x^{(k)}$ is replaced by $x^{(k+1)}$ in each iteration. It tends to be faster than the Jacobi algorithm.

EXAMPLE.– We will solve the system in example [8.2.1]. The Gauss–Seidel iterations are as follows:

$$\begin{cases} x_1^{(k)} = & & \frac{1}{10}x_2^{(k-1)} & - \frac{1}{5}x_3^{(k-1)} & & +\frac{3}{5} \\ x_2^{(k)} = & \frac{1}{11}x_1^{(k)} & & + \frac{1}{11}x_3^{(k-1)} & - \frac{3}{11}x_4^{(k-1)} & +\frac{25}{11} \\ x_3^{(k)} = & -\frac{1}{5}x_1^{(k)} & + \frac{1}{10}x_2^{(k)} & & + \frac{1}{10}x_4^{(k-1)} & -\frac{11}{50} \\ x_4^{(k)} = & & - \frac{3}{8}x_2^{(k)} & + \frac{1}{8}x_3^{(k)} & & +\frac{15}{8} \end{cases}$$

Again, we initiate with $x^{(0)} = (0,0,0,0)^T$ and terminate the method once the following condition is met:

$$\frac{\|x^{(n+1)} - x^{(n)}\|_\infty}{\|x^{(n)}\|_\infty} \leq 10^{-3}.$$

We find that $n = 4$, and more precisely that:

$$\frac{\|x^{(5)} - x^{(4)}\|_\infty}{\|x^{(4)}\|_\infty} = \frac{0.0008}{2.000} = 4 \times 10^{-4},$$

which gives the results shown in Table 8.2.

Thus, we only require five iterations to find the solution almost exactly.

k	$x_1^{(k)}$	$x_2^{(k)}$	$x_3^{(k)}$	$x_4^{(k)}$
0	0.0000	0.0000	0.0000	0.0000
1	0.6000	2.3272	-0.9873	0.8789
2	1.0300	2.0370	–1.0140	0.9844
3	1.0065	2.0036	–1.0025	0.9983
4	1.0009	2.0003	–1.0003	0.9999
5	1.0001	2.0000	–1.000	1.0000

Table 8.2. *Results of each iteration of the Gauss–Seidel algorithm*

EXAMPLE.– Consider the system

$$\begin{cases} 7x & -- & 3y & + & 2z & = 7 \\ 3x & + & 11y & -- & z & = 2 \\ 5x & + & 3y & + & 12z & = 4 \end{cases}$$

We will use this system to compare the Gaussian method to the Gauss–Seidel method, after verifying that these methods are applicable to this problem.

The solution of the system obtained by the Gaussian method is $(x, y, z) = (\frac{953}{976}, -\frac{87}{976}, -\frac{25}{488}) = (0.9764, -0.0891, -0.0512)$.

The Gauss–Seidel method converges because the matrix of the system is strictly diagonally dominant. We have that $x^{(0)} = (0, 0, 0)^T$, and, for $k \geq 0$:

$$\begin{cases} x_1^{(k+1)} & = \frac{1}{7}(7 + 3x_2^{(k)} + 2x_3^{(k)}) \\ x_2^{(k+1)} & = \frac{1}{11}(2 - 3x_1^{(k+1)} + x_3^{(k)}) \\ x_3^{(k+1)} & = \frac{1}{12}(4 - 5x_1^{(k+1)} - 4x_2^{(k+1)}) \end{cases}$$

k	1	2	3	4
$x_1^{(k)}$	1	0.9784	0.9760	0.9764
$x_2^{(k)}$	−0.0909	−0.0905	−0.0891	−0.0891
$x_3^{(k)}$	−0.0606	−0.0517	−0.0511	−0.0512

8.2.3. *Relaxation method*

Suppose that $\omega \in \mathbb{R}^*$. We split the matrix A into $A = D - E - F$ as above. This can be further rewritten as:

$$A = \frac{1}{\omega}(D - \omega E) - \frac{1}{\omega}((1 - \omega)D + \omega F).$$

The matrix B therefore has the expression:

$$B = (D - \omega E)^{-1}((1 - \omega)D + \omega F),$$

and is usually denoted $\mathcal{L}_\omega$. The relaxation method may be stated as follows:

$$\left(\frac{1}{\omega}D - E\right)x^{(k+1)} = \left(\frac{\omega - 1}{\omega}D + F\right)x^{(k)} + b. \qquad [8.19]$$

Its recurrence relation is:

$$\begin{cases} x_1^{(k+1)} = \dfrac{\omega}{a_{11}}\left(-\dfrac{\omega-1}{\omega}a_{11}x_1^{(k)} - \displaystyle\sum_{j>1} a_{1j}x_j^{(k)} + b_1\right) \\ \vdots \\ x_i^{(k+1)} = \dfrac{\omega}{a_{ii}}\left(-\displaystyle\sum_{j<i} a_{ij}x_j^{(k+1)} - \dfrac{\omega-1}{\omega}a_{ii}x_i^{(k)} - \displaystyle\sum_{j>i} a_{ij}x_j^{(k)} + b_i\right) \\ \vdots \\ x_n^{(k+1)} = \dfrac{\omega}{a_{nn}}\left(-\displaystyle\sum_{j<n} a_{nj}x_j^{(k+1)} - \dfrac{\omega-1}{\omega}a_{nn}x_n^{(k)} + b_n\right) \end{cases}$$

We say that ω is the *relaxation factor*:

– if $\omega < 1$, the method is said to be a *sub-relaxation method*;

– if $\omega = 1$, we recover the Gauss–Seidel method;

– if $\omega > 1$, the method is said to be an *over-relaxation method*.

EXAMPLE.– Suppose that we wish to solve the following system using the relaxation method:

$$\begin{cases} 4x_1 + 3x_2 = 24 \\ 3x_1 + 4x_2 - x_3 = 30 \\ 2x_1 - x_2 + 10x_3 - x_4 = -11 \\ - x_2 + 4x_3 = -24 \end{cases} \qquad [8.20]$$

This system has the solution $(3, 4, -5)^T$. We will attempt to solve it using both the Gauss–Seidel method and the relaxation method with a relaxation factor of $\omega = 1.25$.

Both methods are initiated with the vector $x^{(0)} = (1,1,1)^T$. For $k = 1,2,\dots$, the Gauss–Seidel method is as follows:

$$\begin{aligned} x_1^{(k)} &= && -0.75x_2^{(k-1)} && + 6 \\ x_2^{(k)} &= -0.75x_1^{(k)} && && +0.25x_3^{(k-1)} + 7.5 \\ x_3^{(k)} &= && 0.25x_2^{(k)} && - 6 \end{aligned}$$

The relaxation method with $\omega = 1.25$ is as follows:

$$\begin{aligned} x_1^{(k)} &= -0.25x_1^{(k-1)} - 0.9375x_2^{(k-1)} && && + 7.5 \\ x_2^{(k)} &= -0.9375x_1^{(k)} - 0.25x_2^{(k-1)} && + 0.3125x_3^{(k-1)} && + 9.375 \\ x_3^{(k)} &= 0.3125x_2^{(k)} && - 0.25x_3^{(k-1)} && - 7.5 \end{aligned}$$

The results of the first seven iterations of both methods are listed in Tables 8.3 and 8.4 up to seven digits after the decimal point. The Gauss–Seidel method requires 34 iterations to converge, whereas the relaxation method only requires 14 iterations for $\omega = 1.25$.

k	0	1	2	3	4	5	6	7
$x_1^{(k)}$	1	5.250000	3.1406250	3.0878306	3.0549316	3.0343323	3.0214577	3.0134110
$x_2^{(k)}$	1	3.812500	3.8828125	3.9267578	3.9542236	3.9713898	3.9821186	3.9888241
$x_3^{(k)}$	1	−5.046875	−5.0292969	−5.0183105	−5.0114441	−5.0071526	−5.0044703	−5.0027940

Table 8.3. *Results of the first seven iterations of the Gauss–Seidel method*

k	0	1	2	3	4	5	6	7
$x_1^{(k)}$	1	6.3125000	2.6223145	3.1333027	2.9570512	3.0037211	2.9663276	3.0000498
$x_2^{(k)}$	1	3.5295313	3.9585226	4.0102646	4.0074838	4.0029250	4.0009262	4.0002586
$x_3^{(k)}$	1	−6.6501465	−4.6004238	−5.0966863	−4.9734897	−5.0057135	−4.9982822	−5.0003486

Table 8.4. *Results of the first seven iterations of the relaxation method*

8.2.4. *Block forms of the Jacobi, Gauss–Seidel and relaxation methods*

The Jacobi, Gauss–Seidel and relaxation techniques can be applied to matrices divided into blocks.

Simply adapt the formulas of each technique as required.

For example, instead of computing the product of two coefficients as usual, we compute the matrix product of two blocks. Instead of inverting coefficients, we take the matrix inverse of each block. Thus, for each i, the diagonal block A_{ii} must be invertible.

8.3. Convergence of iterative methods

By theorem 8.1, the method defined by the formula:

$$x^{(k+1)} = Bx^{(k)} + g$$

converges if and only if $\rho(B) < 1$.

The problem of determining whether an iterative method converges is therefore an eigenvalue problem, which is usually more difficult to solve than the original problem. Therefore, we shall present a few sufficient conditions below that are simpler to implement.

THEOREM 8.2.– *Let $C \in \mathcal{M}_n(K)$ be a matrix with eigenvalues λ_1, λ_2,..., λ_n. Then:*

$$\{\lambda_1, \cdots, \lambda_n\} \subset \cup_{i=1}^{n} \{z \in \mathbb{C};\ |C(i,j) - z| \leq \sum_{j \neq i} |C(i,j)|\}.$$

The disks with center $C(i,j)$ and radius $\sum_{j \neq i} |C(i,j)|$ in the complex plane are known as Gershgorin circles.

The above proposition implies the following results for the Jacobi and Gauss–Seidel methods:

THEOREM 8.3.– *If A is a strictly diagonally dominant matrix, then the Jacobi and Gauss-Seidel methods are convergent.*

THEOREM 8.4.– *If A is a symmetric positive-definitive matrix, then the Gauss–Seidel method converges (but the Jacobi method does not necessarily converge).*

Note that a matrix A is said to be strictly diagonally dominant if and only if:

$$|a_{ii}| > \sum_{\substack{j=1 \\ j \neq 1}}^{n} |a_{ij}| \quad \text{for } i = 1, \ldots, n.$$

EXAMPLE.– Suppose that we wish to solve the linear system $Ax = b$ using both the Jacobi and Gauss–Seidel methods, where:

$$A = \begin{pmatrix} \alpha & 0 & \gamma \\ 0 & \alpha & \beta \\ 0 & \delta & \alpha \end{pmatrix} \qquad b = \begin{pmatrix} 1 \\ 2 \\ 3 \end{pmatrix},$$

and $\alpha \neq 0$, β, γ and δ are real-valued parameters.

The Jacobi method is as follows:

$$\begin{cases} x_1^{(k+1)} = \dfrac{1}{\alpha}(1 - \gamma x_3^{(k)}) \\ x_2^{(k+1)} = \dfrac{1}{\alpha}(2 - \beta x_3^{(k)}) \\ x_3^{(k+1)} = \dfrac{1}{\alpha}(3 - \delta x_2^{(k)}) \end{cases}$$

The Gauss–Seidel method is as follows:

$$\begin{cases} x_1^{(k+1)} = \dfrac{1}{\alpha}(1 - \gamma x_3^{(k)}) \\ x_2^{(k+1)} = \dfrac{1}{\alpha}(2 - \beta x_3^{(k)}) \\ x_3^{(k+1)} = \dfrac{1}{\alpha}(3 - \delta x_2^{(k+1)}) \end{cases}$$

By decomposing the matrix A, we find:

$$D = \begin{pmatrix} \alpha & 0 & 0 \\ 0 & \alpha & 0 \\ 0 & 0 & \alpha \end{pmatrix}, \quad E = \begin{pmatrix} 0 & 0 & 0 \\ 0 & 0 & 0 \\ 0 & -\delta & 0 \end{pmatrix}, \quad F = \begin{pmatrix} 0 & 0 & -\gamma \\ 0 & 0 & -\beta \\ 0 & 0 & 0 \end{pmatrix},$$

which satisfy $D - A = E + F$.

Therefore, the Jacobi matrix is:

$$B_J = D^{-1}(D - A) = \begin{pmatrix} 0 & 0 & -\dfrac{\gamma}{\alpha} \\ 0 & 0 & -\dfrac{\beta}{\alpha} \\ 0 & -\dfrac{\delta}{\alpha} & 0 \end{pmatrix},$$

and $\det(\lambda I - B_J) = \lambda^3 - \lambda\frac{\beta\delta}{\alpha^2}$.

This converges if and only if $\rho(B_J) < 1$. Therefore, $|\beta\delta| < \alpha^2$.

The Gauss–Seidel matrix is:

$$B_{GS} = (D - E)^{-1}F = \begin{pmatrix} 0 & 0 & -\dfrac{\gamma}{\alpha} \\ 0 & 0 & -\dfrac{\beta}{\alpha} \\ 0 & \dfrac{\delta\beta}{\alpha} & 0 \end{pmatrix},$$

and $\det(\lambda I - B_{GS}) = \lambda(\lambda^2 + \dfrac{\delta\beta^2}{\alpha^2})$.

The necessary and sufficient condition is $\rho(B_{GS}) < 1$. Therefore, $|\beta^2\delta| < \alpha^3$.

THEOREM 8.5.– *Let* $\mathcal{L}_\omega = (D - \omega E)^{-1}((1-\omega)D + \omega F)$, $\omega \neq 0$. *Then:*

$$\rho(\mathcal{L}) \geq |\omega - 1|.$$

It follows that the relaxation method diverges whenever $|\omega - 1| \geq 1$ or, equivalently, whenever $\omega \notin]0, 2[$.

If A is symmetric positive-definite, then the method converges if and only if $0 < \omega < 2$. Furthermore, if A is block-tridiagonal, the following result holds:

THEOREM 8.6.– *Suppose that the eigenvalues of the matrix* B_J *are real and the spectral radius satisfies* $\rho(B_J) < 1$. *Then,* $\rho(\mathcal{L}_\omega)$ *is minimized by the point* ω_0 *on* $]0, 2[$ *(and therefore on* $]0, +\infty[$*), where:*

$$\omega_0 = \frac{2}{1 + \sqrt{1 - [\rho(B_J)]^2}}. \qquad [8.21]$$

REMARK.– Block-tridiagonal matrices play an important role in numerical analysis, because they are often encountered when discretizing linear differential or partial differential problems.

EXAMPLE.– Consider the matrix A of the linear system:

$$A = \begin{pmatrix} 4 & 3 & 0 \\ 3 & 4 & -1 \\ 0 & -1 & 4 \end{pmatrix}.$$

We shall calculate the ω_0 of this matrix. We note that A is a tridiagonal and positive-definite. We have that:

$$B_J = D^{-1}(L+U) = \begin{pmatrix} \frac{1}{4} & 0 & 0 \\ 0 & \frac{1}{4} & 0 \\ 0 & 0 & \frac{1}{4} \end{pmatrix} \begin{pmatrix} 0 & -3 & 0 \\ -3 & 0 & 1 \\ 0 & 1 & 0 \end{pmatrix} \begin{pmatrix} 0 & -0.75 & 0 \\ -0.75 & 0 & 0.25 \\ 0 & 0.25 & 0 \end{pmatrix}.$$

Therefore:

$$\det(B_J - \lambda I) = -\lambda(\lambda^2 - 0.625),$$

and $\rho(B_J) = \sqrt{0.625}$. Hence, the optimal ω_0 is:

$$\omega_0 = \frac{2}{1 + \sqrt{1 - [\rho(B_J)]^2}} = \frac{2}{1 + \sqrt{1 - 0.625}} \approx 1.24.$$

8.4. Conjugate gradient method

Another, even faster, method that can be used whenever A is a symmetric positive-definite matrix is the conjugate gradient method. This time, we write the iterative process as:

$$x^{(k+1)} = x^{(k)} + \alpha_k d^{(k)}, \; k \geq 0, \qquad [8.22]$$

where $d^{(k)} = r^{(k)} - \beta_k d^{(k-1)}$.

Given an initial value $x^{(0)}$, we compute $r^{(0)} = b - Ax^{(0)}$, $d^{(0)} = r^{(0)}$, then, for $k \geq 0$:

$$\alpha_k = \frac{< r^{(k)}, d^{(k)} >}{< Ad^{(k)}, d^{(k)} >}$$
$$x^{(k+1)} = x^{(k)} + \alpha_k d^{(k)}$$
$$r^{(k+1)} = r^{(k)} - \alpha_k Ad^{(k)}$$
$$\beta_{k+1} = \frac{< Ad^{(k)}, r^{(k+1)} >}{< Ad^{(k)}, d^{(k)} >}$$
$$d^{(k+1)} = r^{(k+1)} - \beta_k d^{(k)}.$$

EXAMPLE.– Suppose that we wish to solve the following system using the conjugate gradient method:

$$\begin{cases} 2x_1 - x_2 = 1 \\ -x_1 + 2x_2 = 1 \end{cases} \qquad [8.23]$$

The matrix A and the vector b may be written as:

$$A = \begin{pmatrix} 2 & -1 \\ -1 & 2 \end{pmatrix}, \quad b = \begin{pmatrix} 1 \\ 1 \end{pmatrix}.$$

Note that the matrix A is indeed symmetric. Furthermore, for any number λ:

$$\det(A - \lambda I) = \begin{vmatrix} 2-\lambda & -1 \\ -1 & 2-\lambda \end{vmatrix} = (2-\lambda)^2 - 1$$

and therefore, $\det(A - \lambda I) = 0$ for $\lambda = 1$ and $\lambda = 3$. The eigenvalues of A are therefore real and strictly positive, so the matrix A is symmetric positive-definite.

After computing the iterations, we find:

$$r^0 = b - Ax^0 = \begin{pmatrix} 0 \\ -\frac{3}{2} \end{pmatrix}, \qquad w^1 = -r^0,$$

$$z^1 = Aw^1 = \begin{pmatrix} -\frac{3}{2} \\ 3 \end{pmatrix}, \qquad \alpha^1 = \frac{(r^0)^T w^1}{(w^1)^T z^1} = -\frac{1}{2},$$

$$x^1 = x^0 + \alpha^1 w^1 = \begin{pmatrix} \frac{3}{2} \\ 5/4 \end{pmatrix}, \qquad r^1 = r^0 - \alpha^1 z^1 = \begin{pmatrix} -\frac{3}{4} \\ 0 \end{pmatrix},$$

$$\beta^1 = \frac{(r^1)^T z^1}{(w^1)^T z^1} = \frac{1}{4}, \qquad w^2 = -r^1 + \beta^1 w^1 = \begin{pmatrix} 3/4 \\ \frac{3}{8} \end{pmatrix},$$

$$z^2 = Aw^2 = \begin{pmatrix} \frac{9}{8} \\ 0 \end{pmatrix}, \qquad \alpha^2 = \frac{(r^1)^T w^2}{(w^2)^T z^2} = -\frac{2}{3},$$

$$x^2 = x^1 + \alpha^2 w^2 = \begin{pmatrix} 1 \\ 1 \end{pmatrix},$$

and x^2 is the exact solution of the system [8.23].

8.5. Using *Matlab*

We can reuse the functions presented in Chapter 7.

8.5.1. *Jacobi method*

The following script can be used to solve linear systems with the Jacobi method:

```
function X=jacobi(A,B,P,delta,maxi)
%Inputs - A is an invertible NxN matrix
%        - B is an Nx1 matrix
%        - P is an Nx1 matrix, the initial vector
%        - delta is the tolerance for convergence
%        - maxi is the maximum number of iterations
%Outputs - X is an Nx1 matrix, the iterated vector

N = length(B); for k=1:maxi
  for j=1:N
       X(j)=(B(j)-A(j,[1:j-1,j+1:N])*P([1:j-1,j+1:N]))/A(j,j);
  end
    err=abs(norm(X'-P));
    relerr = err/(norm(X)+eps);
    P=X';
  if(err<delta)|(relerr<delta)
        break
  end
end X=X';
```

8.5.2. *Relaxation method*

Suppose that we wish to solve the following system by the relaxation method with $\omega = 0.2$:

$$\begin{cases} 10x_1 - x_2 + 2x_3 + \quad = 6 \\ -x_1 + 11x_2 - x_3 + 3x_4 = 25 \\ 2x_1 - x_2 + 10x_3 - x_4 = -11 \\ 3x_2 - x_3 + 8x_4 = 15 \end{cases} \qquad [8.24]$$

The exact solution of this system is $x = (1, 2, -1, 1)^T$.

The following *Matlab* script can be used to solve [8.24] by the relaxation method, choosing the parameter value $\omega = 0.2$ and the error $\epsilon = \sum_{i=1}^{n} |x_{n+1}(i) - x_n(i)| = 10^{-10}$:

```
clear all;
clc;
fprintf('General iterative method\n');
n=3000;
a=[10 -1 2 0;-1 11 -1 3;2 -1 10 -1;0 3 -1 8];
y=[6 25 -11 15];
x=zeros(1,4);

w=0.2;  % relaxation factor: 0<w<1

epsilon=1e-10; for k=1:n
    err=0;
    for i=1:4
        s=0;
        xb=x(i);
        for j=1:4
            if i~=j
                s=s+a(i,j)*x(j);
            end
        end
        x(i)=w*(y(i)-s)/a(i,i)+(1-w)*x(i);
        err=err+abs(x(i)-xb);
    end
```

```
        if (err/3<epsilon)
        fprintf('Iteration no.: %d\t Error = %7.2e\n',k,err);
        break;
    end
end x
```

Executing this script returns:

```
General iterative method Iteration no.: 143  Error = 2.73e-010

x =
    1.0000    2.0000   -1.0000    1.0000
```

9

Numerical Methods for Computing Eigenvalues and Eigenvectors

9.1. Introduction

Many of the numerical methods presented in previous chapters assumed that the eigenvalues or the spectral radius of a matrix was known. In general, being able to compute eigenvalues and eigenvectors is essential in every area of science, but this becomes especially important when solving systems of linear differential equations in stability theory, studying the convergence of iterative processes in physics and chemistry (mechanics, circuits, chemical kinetics, Schrödinger equations) and in the field of automation.

THEOREM.– *If A is an $n \times n$ square matrix, then the eigenvalues of A are contained in the union of the n disks $|z - a_{ii}| \leq \sum_{j=1}^{n} |a_{ij}|$.*

LEMMA.– *An upper bound for the largest eigenvalue of the matrix A is therefore given by:*

$$\left\{ |\lambda| \leq \max_{1\leq i\leq n} \left[\sum_{j=1}^{n} |a_{ij}| \right], \max_{1\leq j\leq n} \left[\sum_{i=1}^{n} |a_{ij}| \right] \right\}.$$

EXAMPLE.– Consider the matrix A defined by:

$$A = \begin{pmatrix} 4 & -1 & 1 \\ 1 & 1 & 1 \\ -2 & 0 & -6 \end{pmatrix}.$$

For this matrix, $\det(A - \lambda I) = -\lambda^3 - \lambda^2 + 23\lambda - 26 = P(\lambda)$, so $P(-6) = 16 > 0$, $P(0) = -26 > 0$, $P(2) = 8$, $P(4) = -14$.

The matrix A has three real eigenvalues $\lambda_1 \in]-6,0[$, $\lambda_2 \in]0,2[$ and $\lambda_3 \in]2,4[$. The value of these eigenvalues can, for example, be approximated using Newton's method, which allows us to find $\lambda_1 \approx -5.7685$; $\lambda_2 \approx 1.2992$ and $\lambda_3 \approx 3.4694$.

9.2. Computing $\det(A - \lambda I)$ directly

Earlier, we presented a series of methods for computing the determinant of a matrix after triangulization. These methods can be used to find the characteristic polynomial of the matrix A. We can also use Lagrange interpolation; the characteristic polynomial of any matrix A of order n is of degree n; therefore, given any $(n+1)$ distinct values $\mu_1, \mu_2, \ldots, \mu_{n+1}$, we can compute $v_i = \det(A - \mu_i I)$; $i = 1, \ldots, (n+1)$, and the interpolating polynomial passing through the points $\{(\mu_i, v_i)\}_{i=1,\ldots,(n+1)}$ is identical to the characteristic polynomial of A. We can then find the roots of this polynomial with an iterative method, which allows us to approximate the eigenvalues of A.

EXAMPLE.– Consider a three-story building with the model shown in Figure 9.1. The functions $x_1(t)$ and $x_2(t)$ denote the horizontal displacements of the second and third floors, respectively, with respect to a fixed coordinate system. We write $2m$ for the mass of the second floor, $2k$ for the stiffness of the displacement of the second floor relative to the first floor, m for the mass of the third floor and k for the stiffness of the displacement of the third floor relative to the second floor.

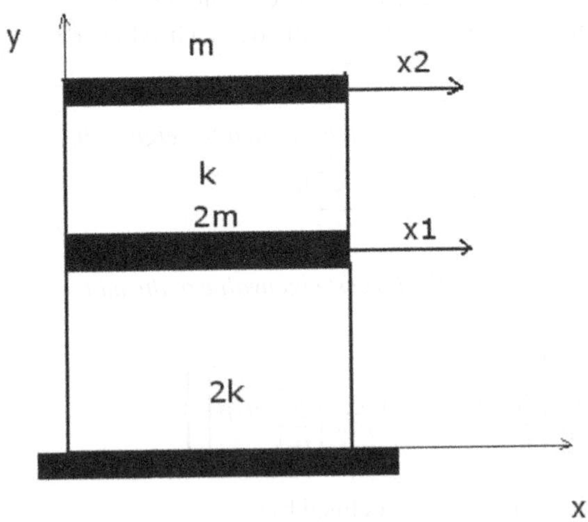

Figure 9.1. *Three-story building*

By applying Newton's second law of motion (the fundamental principle of dynamics), we find the simplified equations of motion (mass-spring):

$$\begin{cases} 2m\ddot{x}_1 + 3kx_1 - kx_2 = 0 \\ m\ddot{x}_2 + kx_2 - kx_1 = 0 \end{cases} \tag{9.1}$$

Equation [9.1] can be rewritten in matrix form as:

$$M\ddot{x}(t) + Kx(t) = 0, \tag{9.2}$$

where:

$$\begin{pmatrix} 2m & 0 \\ 0 & m \end{pmatrix} \begin{pmatrix} \ddot{x}_1 \\ \ddot{x}_2 \end{pmatrix} + \begin{pmatrix} 3k & k \\ -k & k \end{pmatrix} \begin{pmatrix} x_1 \\ x_2 \end{pmatrix} = \begin{pmatrix} 0 \\ 0 \end{pmatrix}.$$

If we look for a solution of the form $x(t) = x_\nu e^{s_\nu t}$, we find that x_ν must satisfy:

$$(K - \lambda_\nu M)x_\nu = 0, \tag{9.3}$$

where $\lambda_\nu = -s_\nu^2$ for $\nu = 1, 2, \ldots, n$.

The matrix M is symmetric positive-definite, so we can decompose it into $M = L^T L$, where L is a regular matrix. System [9.3] then becomes:

$$(KL^{-1} - \lambda_\nu L^T)Lx_\nu = 0.$$

By setting $Y = LX$ and left-multiplying by $(L^T)^{-1}$, we find that:

$$((L^T)^{-1}KV^{-1} - \lambda_\nu I_n)Y = 0.$$

Now, set $A = (L^T)^{-1}KL^{-1}$, giving the matrix system:

$$AY = \lambda Y.$$

We can compute the eigenvalues of this system by solving the equation $\det(K - \lambda_\nu M) = 0$. This yields:

$$\begin{aligned} \det(K - \lambda_\nu M) &= (2k - 2\lambda m)(k - \lambda m) - k^2 \\ &= 2m^2\lambda^2 - 5km\lambda + 2k^2 \\ &= 0. \end{aligned}$$

By setting $\omega_0^2 = \dfrac{k}{m}$, we find that:

$$2\lambda^2 - 5\omega_0^2\lambda + 2\omega_0^4 = 0,$$

which implies that:

$$\lambda_1 = \omega_1^2 = \tfrac{\omega_0^2}{2},$$
$$\lambda_2 = \omega_2^2 = 2\omega_0^2.$$

Next, we compute the eigenvectors:

– For $\lambda_1 = \frac{\omega_0^2}{2}$, by writing $x_1^T = (x_{11}, x_{21})^T$ for the eigenvector, we find:

$$(2k - 2m\lambda_1)x_{11} - kx_{21} = 0 \Rightarrow x_1 = x_{11}\begin{pmatrix} 1 \\ 2 \end{pmatrix}.$$

– For $\lambda_2 = 2\omega_0^2$, by writing $x_2^T = (x_{21}, x_{22})^T$ for the eigenvector, we find:

$$x_2 = x_{22}\begin{pmatrix} 1 \\ -1 \end{pmatrix}.$$

From $Y^T MY = I_2$, we deduce that:

$$x_{1,normalized} = \frac{1}{\sqrt{6m}}\begin{pmatrix} 1 \\ 2 \end{pmatrix} \quad x_{2,normalized} = \frac{1}{\sqrt{3m}}\begin{pmatrix} 1 \\ -1 \end{pmatrix}.$$

REMARK.–

1) In mechanics, the eigenvalues are the natural frequencies of the system. The eigenvectors are the natural modes of the system.

2) In vibrational mechanics, it can be useful to compute the natural frequencies and modes, because the natural frequencies describe the instabilities of the system and the natural modes describe the maximum displacement of each of them.

3) The value ω_0 is the angular frequency of the system.

9.3. Krylov methods

Krylov methods compute the coefficients of the characteristic polynomial by approximating its roots with classical algorithms (e.g. bisection, Newton, sectant).

Let $P(\lambda) = (-1)^n \left(\lambda^n - \sum_{k=1}^{n} a_k \lambda^{n-k} \right)$ be the characteristic polynomial of the matrix A. By the Caley–Hamilton theorem, $P(A) = 0$, so $A^n = \sum_{k=1}^{n} a_k A^{n-k}$.

Suppose that x_0 is an arbitrary non-zero vector. Then:

$$A^n x_0 = \sum_{k=1}^{n} a_k A^{n-k} x_0. \quad [9.4]$$

We write a for the vector with components $(a_i)_{i=1}^n$; $x_1 = Ax_0$, $x_2 = A^2 x_0$, ..., $x_{n-1} = A^{n-1} x_0$ and B for the matrix with columns $x_{n-1}, x_{n-2}, ..., x_0$.

Equation [9.4] can now be rewritten as:

$$A^n x_0 = Ba. \quad [9.5]$$

After choosing an arbitrary x_0, we need to find the solution a of equation [9.5]. If the system is invertible, this allows us to find the coefficients of the characteristic polynomial.

EXAMPLE.– Consider the matrix:

$$A = \begin{pmatrix} 4 & 1 & -1 \\ -6 & -1 & 2 \\ 2 & 1 & 1 \end{pmatrix}.$$

For this matrix, $\det(A - \lambda I) = -\lambda^3 + 4\lambda^2 - 5\lambda + 2 = (\lambda - 1)^2 (2 - \lambda)$.

Choose:

$$x_0 = \begin{pmatrix} 1 \\ 0 \\ 0 \end{pmatrix}, \; A^2 = \begin{pmatrix} 8 & 2 & -3 \\ -14 & -3 & 6 \\ 4 & 2 & 1 \end{pmatrix} \text{ and } A^3 = \begin{pmatrix} 14 & 3 & -7 \\ -26 & -5 & 14 \\ 6 & 3 & 1 \end{pmatrix}.$$

Then, equation [9.5] gives:

$$\begin{pmatrix} 8 & 4 & 1 \\ -14 & -6 & 0 \\ 4 & 2 & 0 \end{pmatrix} \begin{pmatrix} a_1 \\ a_2 \\ a_3 \end{pmatrix} = \begin{pmatrix} 14 \\ -26 \\ 6 \end{pmatrix}$$

$$\Rightarrow \begin{pmatrix} a_1 \\ a_2 \\ a_3 \end{pmatrix} = \begin{pmatrix} 0 & \frac{-1}{2} & \frac{-3}{2} \\ 0 & 1 & \frac{7}{2} \\ 1 & 0 & -2 \end{pmatrix} \begin{pmatrix} 14 \\ -26 \\ 6 \end{pmatrix} = \begin{pmatrix} 4 \\ -5 \\ 2 \end{pmatrix}.$$

9.4. LeVerrier method

The LeVerrier method finds the coefficients of the characteristic polynomial using Newton's formula, which gives a relation between the roots of a polynomial and its coefficients.

We write $Q(x) = a_1 x^n + a_2 x_{n-1} + \cdots + a_n x + a_{n+1}$, where $a_1 \neq 0$.

By Newton's relations:

$$\begin{cases} a_2 + a_1 S_1 & = 0 \\ 2a_3 + a_2 S_1 + a_1 S_2 & = 0 \\ \vdots & \vdots \\ k a_{k+1} + a_k S_1 + \cdots + a_1 S_k & = 0 \\ \vdots & \vdots \\ n a_{n+1} + a_n S_1 + \cdots + a_{n-k+1} S_k + \cdots + a_1 S_n & = 0 \end{cases}$$

where $S_k = \sum_{i=1}^{n} x_i^k$ and x_i is a root of the polynomial Q. Therefore, by considering the characteristic polynomial of A, whose roots are the eigenvalues λ_i of A, we can find $S_k = tr\left(A^k\right)$ and $a_1 = (-1)^n$. This allows us to compute the coefficients a_k for $k = 2, \cdots, n+1$. Specifically:

$$a_k = -\frac{1}{k-1}\left(a_{k-1} S_1 + \cdots + a_2 S_{k-2} + a_1 S_{k-1}\right).$$

REMARK.– One major disadvantage of the LeVerrier method is that it often requires us to compute high powers of the initial matrix A.

EXAMPLE.– Consider the matrix $A = \begin{pmatrix} -1 & -1 & 0 \\ \frac{7}{2} & \frac{3}{2} & -1 \\ \frac{1}{2} & \frac{1}{2} & -1 \end{pmatrix}$.

Then:

$$A^2 = \begin{pmatrix} \frac{-5}{2} & \frac{-1}{2} & 1 \\ \frac{5}{4} & \frac{-7}{4} & \frac{-1}{2} \\ \frac{3}{4} & \frac{-1}{4} & \frac{1}{2} \end{pmatrix},$$

and $\operatorname{diag}\left(A^3\right) = \left(\frac{5}{4}, \frac{-33}{8}, \frac{-1}{4}\right)$, so $a_2 = \operatorname{tr}(A) = \frac{-1}{2}$, $a_3 = -\frac{1}{2}\left(a_2 S_1 + a_1 S_2\right) = -2$ and $a_4 = \frac{-1}{3}\left(a_3 S_1 + a_2 S_2 + a_1 S_3\right) = -2$. Hence, $\det(A - \lambda I) = -\lambda^3 - \frac{1}{2}\lambda^2 - 2\lambda - 2$.

9.5. Jacobi method

The Jacobi method is an iterative method that can be applied whenever the matrix A is symmetric. The Jacobi method involves allowing the group of planar rotations to act upon A, i.e. applying orthogonal transformations to express A in diagonal form. The diagonal elements are then the eigenvalues of the matrix A.

The underlying principle of the method is as follows: consider the matrix H with the same elements as the identity matrix except for the four values $h_{pp} = \cos\alpha$, $h_{pq} = \sin\alpha$, $h_{qq} = \cos\alpha$ and $h_{qp} = -\sin\alpha$, where $p < q$. This matrix H is an orthogonal matrix, i.e. $H^T H = I_n$.

We begin by computing the matrix $A_1 = H_1^{-1} A H = H^T A H$. Note that this only changes the p-th and q-th rows and columns. For $j = p$ or $j = q$, we have that:

$$\begin{cases} a_{pj}^{(1)} = a_{jp}^{(1)} = a_{jp}\cos\alpha - a_{jq}\sin\alpha \\ a_{qj}^{(1)} = a_{jq}^{(1)} = a_{jp}\cos\alpha + a_{jq}\sin\alpha \\ a_{pp}^{(1)} = a_{pp}\cos^2\alpha + a_{qq}\sin^2\alpha - 2\sin\alpha\cos\alpha \\ a_{qq}^{(1)} = a_{qq}\cos^2\alpha + a_{pp}\sin^2\alpha + 2a_{pq}\sin\alpha\cos\alpha \\ a_{pq}^{(1)} = a_{qp}^{(1)} = q_{pq}(\cos^2\alpha - \sin^2\alpha) + (a_{pp} - a_{qq})\sin\alpha\cos\alpha. \end{cases}$$

We can therefore choose α such that $a_{pq}^{(1)} = a_{qp}^{(1)} = 0$. This α satisfies:

$$\tan 2\alpha = \frac{2a_{pq}}{a_{pp} - a_{qq}} \equiv \theta \quad \text{for } |\alpha| \leq \frac{\pi}{4},$$

i.e.:

$$\cos\alpha = \sqrt{\frac{1}{2}\left(1 + \frac{1}{\sqrt{1+\theta^2}}\right)},$$

$$\sin\alpha = \operatorname{sgn}(\theta)\sqrt{\frac{1}{2}\left(1 - \frac{1}{\sqrt{1+\theta^2}}\right)}.$$

If $a_{pp} = a_{qq}$, we set:

$$\cos\alpha = \frac{1}{\sqrt{2}},$$

$$\sin\alpha = \frac{\operatorname{sgn}(a_{pq})}{\sqrt{2}}.$$

We therefore have that:

$$(a_{pp}^{(1)})^2 + (a_{qq}^{(1)})^2 = a_{pp}^2 + a_{qq}^2 + 2a_{pq}^2.$$

As $a_{ii}^{(1)} = a_{ii}$ for $i \neq p$ or q,

$$\sum_{i=1}^{n} (a_{ii}^{(1)})^2 = \sum_{i=1}^{n} a_{ii}^2 + 2a_{pq}^2.$$

Replacing the matrix A by A_1 increases the sum of the squares of the diagonal elements by $2a_{pq}^2$. Iterating this process leads to:

$$A_{k+1} = (H_1 H_2 \dots H_k)^T A (H_1 H_2 \dots H_k).$$

The sequence of matrices A_k converges to a diagonal matrix whose diagonal elements are the eigenvalues of the initial matrix A. The sequence of matrices $P_k = H_1 H_2 \dots H_k$ converges to the matrix whose columns are the eigenvectors of these eigenvalues. Some terms may become non-zero after subsequent iterations, but it can be shown that:

$$\lim_{k \to \infty} \sum_{i \neq j} (a_{ij}^{(k)})^2 = 0.$$

The iterative process is terminated when:

$$1 - \frac{\sum_{i=1}^{n} (a_{ii}^{(k)})^2}{\sum_{i=1}^{n} (a_{ii}^{(k+1)})^2} < \epsilon.$$

When implementing this algorithm, we need a way to choose the pair (p, q) at each iteration. There are several strategies for doing this:

– The classic Jacobi method selects (p, q) such that:

$$|a_{pq}^{(k)}| = \sup_{i \neq j} |a_{ij}^{(k)}|.$$

– The cyclic Jacobi method sweeps through every pair of coefficients by choosing (p, q) as the pairs $(1, 2)$, $(1, 3)$,..., $(1, n)$, then $(2, 3)$,..., $(2, n)$, etc., up to $(n - 1, n)$.

– The cyclic threshold Jacobi method sweeps through the upper triangular elements. Each element a_{ij} is chosen to cancel a_{pq}, but the pair (p, q) is only considered if $|a_{ij}|$ exceeds a certain threshold. The threshold can be adjusted at each iteration.

REMARK.– The Jacobi method is stable, but converges slowly, which means that it is not commonly used in practice.

9.6. Power iteration method

The power iteration method can be used to compute the largest eigenvalue of a diagonalizable matrix. The algorithm is as follows:

$$\begin{cases} \text{Choose arbitrary } q^{(0)} \in \mathbb{R}^n \text{ such that } \left\|q^{(0)}\right\| = 1 \\ x^{(k)} = Aq^{(k-1)},\ k \geq 1 \\ q^{(k)} = \dfrac{x^{(k)}}{\left\|x^{(k)}\right\|}. \end{cases}$$

One essential assumption is required in order for this algorithm to converge: the eigenvalue with the largest modulus must be unique (although this eigenvalue can have multiplicity).

THEOREM.– *Let A be an invertible matrix with one eigenvalue λ_1 whose modulus is strictly greater than the moduli of any of the other eigenvalues. Let q_0 be a vector in $\mathbb{R}^n$ that is not orthogonal to the left eigenspace of λ_1 (i.e. $\langle v_1, q_0\rangle \neq 0$, where v_1 is the eigenvector of λ_1). Then:*

1) the vector $q = \lim_{k\to\infty} \left(\dfrac{|\lambda_1|}{\lambda_1}\right)^k q^{(k)}$ *is an eigenvector of norm* 1 *and eigenvalue* λ_1 $(q = \delta\, v_1)$;

2) $\lim_{k\to\infty} \left\|Aq^{(k)}\right\| = |\lambda_1|$.

EXAMPLE.– Consider the symmetric matrix:

$$A = \begin{pmatrix} 7 & -2 & 0 \\ -2 & 6 & -2 \\ 0 & -2 & 5 \end{pmatrix}.$$

Suppose that we wish to find the eigenvalue λ_1 with the greatest absolute value and the corresponding normalized eigenvector x_1 using the power iteration method with initial vector $v = (1, 0, 0)^T$.

As $|\lambda_1| = \lim_{k\to+\infty} \dfrac{\|A^k v\|}{\|A^{k-1}v\|}$, we need to compute the vectors $A^k v$, $k = 0, 1, \ldots, 6$.

v	Av	A^2v	A^3v	A^4v	A^5v	A^6v
1	7	53	423	3501	29727	257013
0	–2	–26	–270	–2610	–24462	–225666
0	0	4	72	900	9720	97524

We deduce that:

$\frac{\|A^2v\|}{\|Av\|}$	$\frac{\|A^3v\|}{\|A^2v\|}$	$\frac{\|A^4v\|}{\|A^3v\|}$	$\frac{\|A^5v\|}{\|A^4v\|}$	$\frac{\|A^6v\|}{\|A^5v\|}$
8.1275	8.5680	8.7947	8.9055	8.9573

which implies that $|\lambda_1| \approx 8.9573$.

As the components of the vectors $A^k v$ have the same sign, $\lambda_1 \approx 8.9573$. If we had continued the power iteration algorithm, we would have found that $\lambda_1 = 9$. We can verify that this value is correct by substituting it into the characteristic polynomial of the matrix A.

The corresponding normalized eigenvector is $x_1 = \lim_{k\to+\infty} \frac{A^k v}{\|A^k v\|}$, which is

$$x_1 \approx \frac{A^6 v}{\|A^6 v\|} = \begin{pmatrix} 0.7226 \\ -0.6345 \\ 0.2742 \end{pmatrix}.$$

Solving the equation $(A - 9I)x = 0$ gives:

$$x_1 = \begin{pmatrix} \frac{2}{3} \\ -\frac{2}{3} \\ \frac{1}{3} \end{pmatrix}.$$

We can now compute the other eigenvalues λ_2, λ_3 and the corresponding eigenvectors x_2, x_3 analytically.

The characteristic polynomial of A is:

$$\det(A - \lambda I) = -\lambda^3 + 18\lambda^2 - 99\lambda + 162.$$

As $\lambda_1 = 9$, this implies that:

$$\det(A - \lambda I) = (\lambda - 9)(-\lambda^2 + 9\lambda - 18) = -(\lambda - 9)(\lambda - 6)(\lambda - 3),$$

and so $\lambda_2 = 6$ and $\lambda_3 = 3$. Solving the equation $(A - \lambda_i I)x_i = 0$ for $i = 2, 3$ yields the normalized eigenvectors of λ_2 and λ_3:

$$x_2 = \begin{pmatrix} \frac{2}{3} \\ \frac{1}{3} \\ -\frac{2}{3} \end{pmatrix}, \qquad x_3 = \begin{pmatrix} \frac{1}{3} \\ \frac{2}{3} \\ \frac{2}{3} \end{pmatrix}.$$

9.6.1. *Deflation algorithm*

This algorithm can be used to compute the other eigenvalues of A. The idea is as follows: once λ_1 has been found, we construct a matrix A_1 with the eigenvalues 0, λ_2,..., λ_n. We can then find λ_2, etc., providing that these eigenvalues all have distinct modulus values.

As λ_1 is also an eigenvalue of A^T with eigenvector y, we define:

$$A_1 = A - \frac{\lambda_1}{y^T x^{(1)}} x^{(1)} y^T.$$

9.7. Inverse power method

If the matrix A is invertible, the eigenvalues of A^{-1} are the inverses of the eigenvalues of A. We can therefore use the power iteration method to compute the eigenvalue with the smallest modulus by applying it to A^{-1}. We do not need to compute the inverse for this. Instead, we can find the LU factorization of A before performing the rest of the algorithm.

The algorithm proceeds in the same way as the power iteration method, except that we use A instead of A^{-1} when constructing the sequence of intermediate vectors $A^{-1}q^{(k)} = x^{(k+1)}$. Multiplying this equation throughout by A, we find $Ax^{(k+1)} = q^{(k)}$. This is a linear system that we can solve using the LU decomposition of A. Once we have found $x^{(k+1)}$, we continue by renormalizing as before to compute the next iteration $q^{(k+1)}$. The sequence, formed by the iterations of the algorithm, will converge to the target eigenvector.

EXAMPLE.– Consider the diagonal matrix A defined by:

$$A = \begin{pmatrix} 2 & 0 \\ 0 & 1 \end{pmatrix}.$$

We will apply the inverse power iteration algorithm:

$$\lambda^{(n-1)} = \frac{x^{(n-1)^T} A x^{(n-1)}}{||x^{(n-1)}||^2},$$

$$x^{(n)} = (A - \lambda^{(n-1)} I)^{-1} x^{(n-1)}.$$

The initial vector $x^{(0)}$ is chosen to be:

$$x^{(0)} = \begin{pmatrix} x_1^{(0)} \\ x_2^{(0)} \end{pmatrix}.$$

If $0 < x_1^{(0)} < x_2^{(0)}$, then $\lim_{n\to\infty} \lambda^{(n)} = 1$.

We write $x_1^{(n-1)}, x_2^{(n-1)}$ for the components of the vector $x^{(n-1)}$ and assume that:

$$x_1^{(n-1)} \neq 0 \quad \text{and} \quad x_2^{(n-1)} \neq 0$$

(this hypothesis holds for $n = 1$). Then:

$$\lambda^{(n-1)} = \frac{2(x_1^{(n-1)})^2 + (x_2^{(n-1)})^2}{(x_1^{(n-1)})^2 + (x_2^{(n-1)})^2} = \frac{2\left(\dfrac{x_1^{(n-1)}}{x_2^{(n-1)}}\right)^2 + 1}{\left(\dfrac{x_1^{(n-1)}}{x_2^{(n-1)}}\right)^2 + 1}.$$

We also have the system:

$$\begin{pmatrix} 2-\lambda^{(n-1)} & 0 \\ 0 & 1-\lambda^{(n-1)} \end{pmatrix} \begin{pmatrix} x_1^{(n)} \\ x_2^{(n)} \end{pmatrix} = \begin{pmatrix} x_1^{(n-1)} \\ x_2^{(n-1)} \end{pmatrix}.$$

Note that $\lambda^{(n-1)}$ is distinct from 1 and 2. We can therefore solve the linear system to find:

$$\frac{x_1^{(n)}}{x_2^{(n)}} = \frac{x_1^{(n-1)}}{x_2^{(n-1)}} \cdot \frac{1-\lambda^{(n-1)}}{2-\lambda^{(n-1)}}.$$

Finally, we find that:

$$\frac{x_1^{(n)}}{x_2^{(n)}} = -\left(\frac{x_1^{(n-1)}}{x_2^{(n-1)}}\right)^3.$$

We deduce that $x_1^{(n)} \neq 0$ and $x_2^{(n)} \neq 0$, so the algorithm is well defined. Furthermore:

$$\frac{x_1^{(n)}}{x_2^{(n)}} = -\left(\frac{x_1^{(n-1)}}{x_2^{(n-1)}}\right)^3 = \ldots = (-1)^n \left(\frac{x_1^{(0)}}{x_2^{(0)}}\right)^{3^n}.$$

We conclude that:

$$\text{If } 0 < x_1^{(0)} < x_2^{(0)} \text{ then } \lim_{n\to\infty} \frac{x_1^{(n)}}{x_2^{(n)}} = 0 \text{ and } \lim_{n\to\infty} \lambda^{(n)} = 1.$$

Similarly:

$$\text{If } 0 < x_2^{(0)} < x_1^{(0)} \text{ then } \lim_{n\to\infty} \frac{x_1^{(n)}}{x_2^{(n)}} = +\infty \text{ and } \lim_{n\to\infty} \lambda^{(n)} = 2.$$

9.8. Givens–Householder method

The Givens–Householder method is a combination of two algorithms. The Householder method transforms the original matrix A into symmetric tridiagonal form. The Givens method then computes the eigenvalues of this symmetric tridiagonal matrix. Suppose that A has been transformed into the form:

$$B = \begin{pmatrix} b_1 & c_1 & 0 & \cdots & 0 \\ c_1 & b_2 & c_2 & \ddots & \vdots \\ 0 & c_2 & \ddots & \ddots & 0 \\ \vdots & \ddots & \ddots & \ddots & c_{n-1} \\ 0 & \cdots & 0 & c_{n-1} & b_n \end{pmatrix}$$

and we write B_k for the submatrix:

$$B_k = \begin{pmatrix} b_1 & c_1 & 0 & \cdots & 0 \\ c_1 & b_2 & c_2 & \ddots & \vdots \\ 0 & c_2 & \ddots & \ddots & 0 \\ \vdots & \ddots & \ddots & \ddots & c_{k-1} \\ 0 & \cdots & 0 & c_{k-1} & b_k \end{pmatrix}.$$

The characteristic polynomials of the matrices B_k satisfy the recurrence relations:

$$\begin{aligned} p_0(\lambda) &= 1 \\ p_1(\lambda) &= b_1 - \lambda \\ p_k(\lambda) &= (b_k - \lambda)p_{k-1}(\lambda) - c_{k-1}^2 p_{k-2}(\lambda) \quad \text{for } k = 2, \cdots, n. \end{aligned}$$

These polynomials satisfy the following properties:

$$\lim_{\lambda \to -\infty} p_k(\lambda) = +\infty;$$

If $p_k(\lambda_0) = 0$, then $p_{k-1}(\lambda_0)p_{k+1}(\lambda_0) < 0$ for $k = 1, \ldots, n-1$. The polynomial p_k has k distinct real roots that separate the $(k+1)$ roots of the polynomial p_{k+1} (i.e. $x < y < z$, where $p_{k+1}(x) = p_{k+1}(z) = 0$ and $p_k(y) = 0$).

Let a be an arbitrary real number. If we define:

$$\operatorname{sgn}(p_k(a)) = \begin{cases} \operatorname{sgn}(p_k(a)) & \text{if } p_k(a) \neq 0 \\ \operatorname{sgn}(p_{k-1}(a)) & \text{if } p_k(a) = 0 \end{cases}$$

then it can be shown that the number of sign changes $N(k, a)$ between the elements of the ordered set $\{+, \operatorname{sgn}(p_1(a)), \ldots, \operatorname{sgn}(p_k(a))\}$ is equal to the number of roots of the polynomial p_k that are strictly less than a.

9.8.1. *Givens algorithm*

To find an eigenvalue of the matrix B, we choose an arbitrary interval $[a_0, b_0]$ that contains λ_i. Let c_0 be the midpoint of the interval $[a_0, b_0]$:

If $N(n, c_0) \geq i$, then $\lambda_i \in [a_0, c_0[$

If $N(n, c_0) < i$, then $\lambda_i \in [c_0, b_0]$

This restricts the target interval containing λ_i to $[a_1, b_1]$. Repeating this process, we can construct a sequence of nested intervals $[a_k, b_k]$ of size $(b_0 - a_0)/2^k$, each of which contains λ_i.

9.9. Using *Matlab*

We can use the *eig* function to return the eigenvalues of any given matrix in *Matlab*. For example, suppose that we wish to compute the eigenvalues and eigenvectors of the matrix:

$$A = \begin{pmatrix} 0 & -6 & -1 \\ 6 & 2 & -16 \\ -5 & 20 & -10 \end{pmatrix}.$$

We can do this in two different ways:

```
>> A = [0 -6 -1; 6 2 -16; -5 20 -10];
>> lambda = eig(A)

lambda =
        -3.0710
        -2.4645 +17.6008i
        -2.4645 -17.6008i

>> [V,D] = eig(A)

V =
      -0.8326                 0.2003 - 0.1394i    0.2003 + 0.1394i
      -0.3553                -0.2110 - 0.6447i   -0.2110 + 0.6447i
      -0.4248                -0.6930             -0.6930

D =
    -3.0710                  0                    0
          0          -2.4645+17.6008i             0
          0                  0            -2.4645-17.6008i
```

The matrix V contains the eigenvectors, and the matrix D contains the eigenvalues along the diagonal.

Alternatively, we can compute the characteristic polynomial of the matrix A using the *poly(A)* command, and then find the roots of this polynomial with the *roots(poly(A))* command. For example:

```
>> A = [0 -6 -1; 6 2 -16; -5 20 -10];
>> poly(A)

ans =
     1.0000    8.0000  331.0000  970.0000

>> roots(poly(A))
```

```
ans =
    -2.4645 +17.6008i
    -2.4645 -17.6008i
    -3.0710
```

The *eig* command is usually the best way to compute the eigenvalues in *Matlab*.

9.9.1. *Application to a buckling beam*

Consider the problem of a beam that buckles compressively under an axial force P. The displacement field $u(x)$ of the beam satisfies the differential equation:

$$u^{(4)} + \frac{P}{EI}u'' = 0, \qquad [9.6]$$

where E is Young's modulus and I is the moment of inertia of the beam. The boundary conditions are given by:

$$u(0) = u''(0) = 0 \quad u(L) = u'(L) = 0. \qquad [9.7]$$

We want to find answers to the following questions:

1) Using the finite differences method (see the chapter on numerical differentiation), show that analyzing the displacement field is equivalent to solving an eigenvalue problem.

2) Using the Jacobi method, calculate the three first frequencies of the system and the corresponding eigenvectors.

We partition the beam into $n + 1$ segments of length $L/(n + 1)$ (see Figure 9.2).

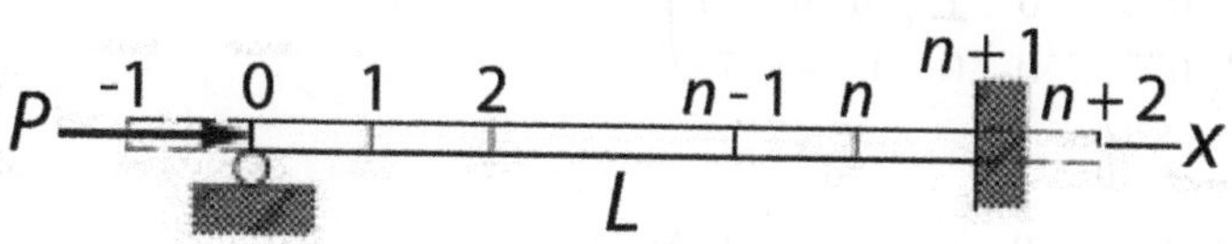

Figure 9.2. *Discretization of the beam*

The derivatives of u in equation [9.6] can be approximated by central divided differences at the interior points (node 1 to n). This gives:

$$\frac{u_{i-2} - 4u_{i-1} + 6u_i - 4u_{i+1} + u_{i+2}}{h^4}$$
$$= \frac{P}{EI}, \frac{-u_{i-1} + 2u_i - u_{i+1}}{h^2}, \quad i = 1, 2, \ldots, n.$$

Multiplying throughout by h^4 gives the system of equations:

$$\begin{aligned} u_{-1} - 4u_0 + 6u_1 - 4u_2 + u_3 &= \lambda(-u_0 + 2u_1 - u_2) \\ u_0 - 4u_1 + 6u_2 - 4u_3 + u_4 &= \lambda(-u_1 + 2u_2 - u_2) \\ &\vdots \\ u_{n-3} - 4u_{n-2} + 6u_{n-1} - 4u_n + u_{n+1} &= \lambda(-u_{n-2} + 2u_{n-1} - u_n) \\ u_{n-2} - 4u_{n-1} + 6u_n - 4u_{n+1} + u_{n+2} &= \lambda(-u_{n-1} + 2u_n - u_{n+1}) \end{aligned} \qquad [9.8]$$

where:

$$\lambda = \frac{Ph^2}{EI} = \frac{PL^2}{(n+1)^2 EI}.$$

The displacements u_{-1}, u_0, u_{n+1} and u_{n+2} can be eliminated using the boundary conditions, leading to:

$$u_0 = 0 \quad u_{-1} = -u_1 \quad u_{n+1} = 0 \quad u_{n+2} = u_n.$$

By writing the system [9.8] in matrix form after performing the necessary substitutions, we can reduce it to an equation of type $Ax = \lambda Bx$, where:

$$A = \begin{pmatrix} 5 & -4 & 1 & 0 & 0 & \dots & 0 \\ -4 & 6 & -4 & 1 & 0 & \dots & 0 \\ 1 & -4 & 6 & -4 & 1 & \dots & 0 \\ \vdots & \ddots & \ddots & \ddots & \ddots & \ddots & \vdots \\ 0 & \dots & 1 & -4 & 6 & -4 & 1 \\ 0 & \dots & 0 & 1 & -4 & 6 & -4 \\ 0 & \dots & 0 & 0 & 1 & -4 & 7 \end{pmatrix}$$

and:

$$B = \begin{pmatrix} 2 & -1 & 0 & 0 & 0 & \dots & 0 \\ -1 & 2 & -1 & 0 & 0 & \dots & 0 \\ 0 & -1 & 2 & -1 & 0 & \dots & 0 \\ 0 & -1 & 2 & -1 & 0 & \dots & 0 \\ \vdots & \ddots & \ddots & \ddots & \ddots & \ddots & \vdots \\ 0 & \dots & 0 & -1 & 2 & -1 & 0 \\ 0 & \dots & 0 & 0 & -1 & 2 & -1 \\ 0 & \dots & 0 & 0 & 0 & -1 & 2 \end{pmatrix}.$$

To solve the question, we can write a *Matlab* script that implements the Jacobi method. The code of the script is as follows:

```
n = 10; A = zeros(n); B = zeros(n);

for i = 1:n
  A(i,i) = 6;
  B(i,i)= 2;
end
A(1,1) = 5;
A(n,n) = 7;
for i = 1:n-1
  A(i,i+1) = -4;
  A(i+1,i) = -4;
  B(i,i+1) = -1;
  B(i+1,i) = -1;
end
for i = 1:n-2
  A(i,i+2) = 1;
  A(i+2,i) = 1;
end [H,T] = stdForm(A,B);
[eVals,Z] = jacobi(H);
X = T*Z;

for i = 1:n
  xMag = sqrt(dot(X(:,i),X(:,i)));
  X(:,i) = X(:,i)/xMag;
end [eVals,X] = sortEigen(eVals,X);
eigenvalues = eVals(1:3)'
eigenvectors = X(:,1:3)
```

Running this script returns the following result:

```
>> eigenvalues =
         0.1641 0.4720 0.9022

eigenvectors =
         0.1641 -0.1848 0.3070
         0.3062 -0.2682 0.3640
         0.4079 -0.1968 0.1467
         0.4574 0.0099 -0.1219
         0.4515 0.2685 -0.1725
         0.3961 0.4711  0.0677
         0.3052 0.5361  0.4089
         0.1986 0.4471 0.5704
         0.0988 0.2602 0.4334
         0.0270 0.0778 0.1486
```

Figure 9.3 shows the first three modes computed by this script.

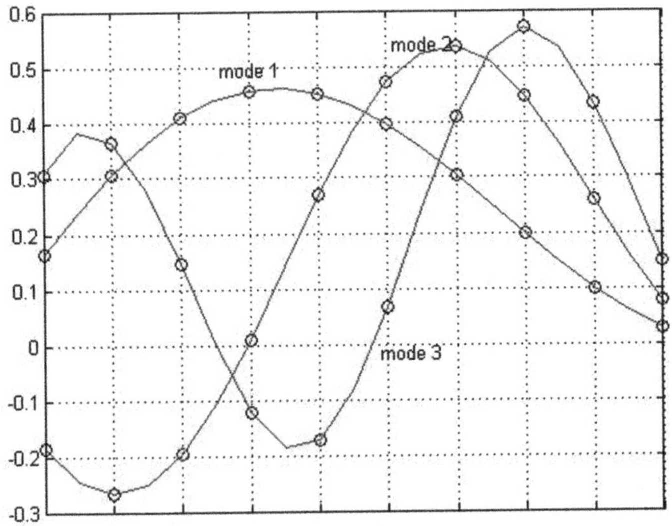

Figure 9.3. *The first modes calculated by the script*

The above script uses the following functions:

```
function [eVals,eVecs] = jacobi(A,tol)
% Jacobi method to compute the eigenvalues
% and eigenvectors of a symmetric matrix A.
% USE: [eVals,eVecs] = jacobi(A,tol)
% tol = error tolerance (set to 1.0e-9 by default).
if nargin < 2;
tol = 1.0e-9;
end
n = size(A,1);
maxRot = 5*(n^2);  % Maximum number of rotations
P = eye(n);        % Initialize the rotation matrix
for i = 1:maxRot   % Begin the rotations
[Amax,k,L] = maxElem(A);
if Amax < tol;
eVals = diag(A);
eVecs = P;
return
end
```

```
[A,P] = rotate(A,P,k,L);
end

error('Too many Jacobi rotations')
```

The *maxElem* function can be implemented as follows:

```
function [Amax,k,L] = maxElem(A)
% Find Amax = A(k,L).
n = size(A,1);
Amax = 0;
for i = 1:n-1
for j = i+1:n

if abs(A(i,j))>= Amax
 Amax = abs(A(i,j));
 k = i;
 L = j;
end
end
end
```

The *rotate* function can be implemented as follows:

```
function [A,P] = rotate(A,P,k,L)
% Zeros the coefficient A(k,L) using Jacobi rotations and
% updates the transformation matrix P.
n = size(A,1);
diff = A(L,L) - A(k,k);

if abs(A(k,L)) < abs(diff)*1.0e-36
t = A(k,L);
else

phi = diff/(2*A(k,L));
t = 1/(abs(phi) + sqrt(phi^2 + 1));

if phi < 0; t = -t; end; end
c = 1/sqrt(t^2 + 1);

s = t*c; tau = s/(1 + c);
temp = A(k,L); A(k,L) = 0;

A(k,k) = A(k,k) - t*temp;
```

```
A(L,L) = A(L,L) + t*temp;
for i = 1:k-1
temp = A(i,k);

A(i,k) = temp -s*(A(i,L) + tau*temp);

A(i,L) = A(i,L) + s*(temp - tau*A(i,L));
end
for i = k+1:L-1

temp = A(k,i);
A(k,i) = temp - s*(A(i,L) + tau*A(k,i));

A(i,L) = A(i,L) + s*(temp - tau*A(i,L));
end
for i = L+1:n

temp = A(k,i);
A(k,i) = temp - s*(A(L,i) + tau*temp);

A(L,i) = A(L,i) + s*(temp - tau*A(L,i));
end
for i = 1:n

temp = P(i,k);
P(i,k) = temp - s*(P(i,L) + tau*P(i,k));

P(i,L) = P(i,L) + s*(temp - tau*P(i,L));
end
```

The *sortEigen* function can be implemented as follows:

```
function [eVals,eVecs] = sortEigen(eVals,eVecs)
% USE: [eVals,eVecs] = sortEigen(eVals,eVecs)
n = length(eVals);
for i = 1:n-1
index = i;
val = eVals(i);

for j = i+1:n
if eVals(j) < val
index = j;
val = eVals(j);
end
```

```
end
if index ~= i
eVals = swapRows(eVals,i,index);

eVecs = swapCols(eVecs,i,index);
end
end
```

The *stdForm* function can be implemented as follows:

```
function [H,T] = stdForm(A,B)
% Transform A*x = lambda*B*x into H*z = lambda*z
% and compute the transformation matrix T such that x = T*z.
% USE: [H,T] = stdForm(A,B)
n = size(A,1);
L = cholesky(B);
Linv = invert(L);

H = Linv*(A*Linv');
T = Linv';
```

The *swapRows* function can be implemented as follows:

```
function v = swapRows(v,i,j)
% Swap the i-th and j-th rows of the vector or matrix v.
temp = v(i,:);
v(i,:) = v(j,:);
v(j,:) = temp;
```

The functions *invert* and *cholesky* were defined earlier in Chapter 7.

10

Least-squares Approximation

10.1. Introduction

One of the most common problems in experimental contexts is to find an analytic formula describing an unknown function f, given known experimental values $f_i = f(x_i)$ on a set of points x_i. Polynomial interpolation can be used, but the results tend to be unsatisfactory. Another idea is to approximate the curve passing through two sets of measurements with a polynomial. The optimal coefficients of the polynomial are those which minimize the variance of the interpolation error. This principle is known as the method of least-squares. In this chapter, we will restrict attention to the discrete case (see Appendix 2).

10.2. Analytic formulation

The problem may be stated as follows:

– Let $((t_1, y_1), \dots, (t_m, y_m))$ be a family of known points in the plane ($(t_i, y_i) \in \mathbb{R}^2$). We shall assume that t_i are distinct.

– Let $(\phi_1 1(t), \dots, \phi_n(t))$, ($\phi_i \colon \mathbb{R} \to \mathbb{R}$) for $n \leq m$ be a linearly independent family of functions. These functions can, for example, be polynomials, exponential functions, rational functions, etc.

– Given the family $x_1, \dots, x_n$ of real numbers, we construct the function ϕ_x by defining:

$$\phi_x(t) = \sum_{i=1}^{n} x_i \phi_i(t). \qquad [10.1]$$

In the case where $m \geq n$, we want to find the family $\hat{x} = (\hat{x}_1, \ldots, \hat{x}_n)$ that minimizes the error:

$$E(x) = \sum_{i=1}^{n} \|y_i \phi_x(t_i)\|^2. \qquad [10.2]$$

In other words, we need to solve the following optimization problem:

$$\text{Find } x \in \mathbb{R}^n, \text{ such that } E(\hat{x}) \leq E(x) \quad \forall x \in \mathbb{R}^n. \qquad [10.3]$$

EXAMPLE.– Consider the set of points $(t_i, y_i) \in \mathbb{R} \times \mathbb{R}$, $i \in \{1, \ldots, m\}$. Suppose that we wish to approximate the scatter plot of these points with a line $\phi(t) = \alpha t + \beta$. With the above notation, $x_1 = \alpha$, $x_2 = \beta$, $\phi_1(t) = t$ and $\phi_2(t) = 1$. The function $E(\alpha, \beta)$ represents the approximation error (in the sense of least-squares) committed when choosing the line defined by α and β. This error can be written as:

$$E(\alpha, \beta) = \sum_{i=1}^{m} (\alpha t_i + \beta - y_i)^2. \qquad [10.4]$$

Let $Y = (y_1, \ldots, y_m)^t$, $X = (x_1, \ldots, x_n)^t$, and suppose that:

$$A = \begin{pmatrix} \phi_1(t_1) & \ldots & \phi_n(t_1) \\ \vdots & & \vdots \\ \phi_1(t_m) & \cdots & \phi_m \end{pmatrix},$$

i.e. $a_{ij} = \phi_j(t_i)$. Clearly:

$$\begin{pmatrix} \phi(t_1) \\ \vdots \\ \phi(t_m) \end{pmatrix} = AX.$$

Therefore, the function $E\colon \mathbb{R}^n \to \mathbb{R}$ can be written in matrix form as:

$$E(X) = \|Y - AX\|_2^2 = X^T A^T AX - 2Y^T AX + Y^T X. \qquad [10.5]$$

DEFINITION.– *Let $S \in \mathcal{M}_{nn}(\mathbb{R})$ be a symmetric matrix. Suppose that $v \in \mathbb{R}^n$. The function $f\colon \mathbb{R}^n \to \mathbb{R}$ defined by:*

$$f(X) =< SX|X > + < X|v >= X^T SX + v^T X \qquad [10.6]$$

is said to be a quadratic function.

Let f be a function from $\mathbb{R}^n$ to $\mathbb{R}$, and write $\nabla f(X)$ be the gradient of f at X. As we will see later, a necessary condition for $\hat{X}$ to be a solution of:

$$f(\hat{X}) \leq f(X) \quad \forall X \in \mathbb{R}^n$$

which is given by:

$$\nabla f(\hat{X}) = 0.$$

For the quadratic function [10.6], the gradient can be written in the following form:

$$\nabla f(X) = 2X^T S + v^T. \qquad [10.7]$$

THEOREM.– *A necessary condition for the vector $\hat{X} \in \mathbb{R}^n$ to be a solution of the least-squares problem [10.3] is given by the relation:*

$$A^T A\hat{X} = A^T Y. \qquad [10.8]$$

DEFINITION.– *The relations [10.8] are said to be the normal equations of the least-squares problem [10.3].*

EXAMPLE.– Suppose that we wish to find an expression of the form $P(x) = A \exp Mx$ for the data points:

x_i	1	2	3	4
P_i	7	11	17	27

To do this, we set $y = \ln P$ and $B = \ln A$, so that $y = \ln P = \ln A + Mx$. We will therefore be working with the data points $(x_i, y_i) = (x_i, \ln P_i)$:

x_i	1	2	3	4
y_i	1.95	2.4	2.83	3.3

We now simply need to solve the system of normal equations:

$$\begin{cases} s_0 B + s_1 M = t_0 \\ s_1 B + s_2 M = t_1 \end{cases}$$

where:

$$s_0 = 4 \; s_k = \sum_{i=1}^{4} x_i^k \; k = 1, 2$$

$$t_0 = \sum_{i=1}^{4} y_i \qquad t_1 = \sum_{i=1}^{4} y_i x_i$$

which gives:

$$s_0 = 4, \; s_1 = 10, \; s_2 = 30, \; t_0 = 10.48 \text{ and } t_1 = 128.44.$$

These values imply that $M \simeq 0.45$, $B \simeq 1.5$ and:

$$P(x) = 4.48e^{0.45x}.$$

REMARK.– In the case where rank$(A) = n$, i.e. the columns of A are linearly independent, the matrix A^TA is symmetric positive-definite, and therefore it is always invertible, so the equations [10.8] have a unique solution.

REMARK.– Case $m = n$; Suppose that $m = n$ and A has rank n. Then, A is invertible (and so is A^T). The system [10.8] is equivalent to $A\hat{X} = Y$. Therefore:

$$\forall i \in \{1, \cdots, m\},\ \hat{\phi}(t_i) = y_i \text{ where } \hat{\phi} = \sum_{i=1}^{n} \hat{x}_i \phi_i.$$

This case therefore reduces to polynomial interpolation.

EXAMPLE.– Suppose that we wish to find the polynomial $p(x) = a_0 + a_1 x + \cdots + a_m x^m$ that is minimized by:

$$S = \sum_{i=0}^{N} (y_i - a_0 - a_1 x_i - \cdots - a_m x_i^m)^2 .$$

The data points (x_i, y_i) are known and $m < N$.

We can restate the first condition for a minimum as:

$$\frac{\partial S}{\partial a_k} = -2 \sum_{i=0}^{N} x_i^k [y_i - a_0 - a_1 x_i - \cdots - a_m x_i^m] = 0, \qquad [10.9]$$

where $k = 0, 1, \ldots, m$.

By introducing the symbols $s_k = \displaystyle\sum_{i=0}^{N} x_i^k$ and $t_k = \sum_{i=0}^{N} y_i x_i^k$, equation [10.9] may be rewritten as:

$$\begin{cases} t_0 &= s_0 a_0 &+ s_1 a_1 & \ldots & s_m a_m \\ t_1 &= s_1 a_0 &+ s_2 a_1 & \ldots & s_{m+1} a_m \\ \vdots &= \vdots &+ \vdots & \vdots & \vdots \\ t_m &= s_m a_0 &+ s_{m+1} a_1 & \ldots & s_{2m} a_m \end{cases} \qquad [10.10]$$

These equations are known as the normal equations.

We will begin by showing that the determinants:

$$D_k = \begin{vmatrix} s_0 & s_1 & \ldots & s_{k-1} \\ s_1 & s_2 & \ldots & s_k \\ \vdots & \vdots & \ddots & \vdots \\ s_{k-1} & s_k & \ldots & s_{2k-2} \end{vmatrix}$$

are strictly positive for $k = 1, 2, \ldots, N + 1$.

We set:

$$P_k(x) = \begin{vmatrix} s_0 & s_1 & \dots & s_k \\ s_1 & s_2 & \dots & s_{k+1} \\ \vdots & \vdots & \ddots & \vdots \\ s_{k-1} & s_k & \dots & s_{2k-1} \\ 1 & x & \dots & x^k \end{vmatrix}.$$

Next, we multiply the last row of $P_k(x)$ by x^j and then sum over i. For $j < k$:

$$\sum_{i=0}^{N} x_i^j P_k(x_i) = \begin{vmatrix} s_0 & s_1 & \dots & s_k \\ s_1 & s_2 & \dots & s_{k+1} \\ \vdots & \vdots & \ddots & \vdots \\ s_{k-1} & s_k & \dots & s_{2k-1} \\ s_j & s_{j+1} & \dots & s_{j+k} \end{vmatrix} = 0,$$

as two of the rows are identical. For $j = k$:

$$\sum_{i=0}^{N} x_i^k P_k(x_i) = D_{k+1}.$$

Now, multiplying the bottom row of $P_k(x)$ by $P_k(x)$ and summing over i, we find that:

$$S_k = \sum_{i=0}^{N} P_k(x_i)^2 = \begin{vmatrix} s_0 & s_1 & \dots & s_k \\ s_1 & s_2 & \dots & s_{k+1} \\ \vdots & \vdots & \ddots & \vdots \\ s_{k-1} & s_k & \dots & s_{2k-1} \\ \sum_{i=0}^{N} P_k(x_i) & \sum_{i=0}^{N} x_i P_k(x_i) & \dots & \sum_{i=0}^{N} x_i^k P_k(x_i) \end{vmatrix}$$

$$= \begin{vmatrix} s_0 & s_1 & \dots & s_k \\ s_1 & s_2 & \dots & s_{k+1} \\ \vdots & \vdots & \ddots & \vdots \\ s_{k-1} & s_k & \dots & s_{2k-1} \\ 0 & 0 & \dots & P_{k+1}(x_i) \end{vmatrix}$$

$$= D_{k+1} D_k.$$

Therefore:

$$S_k = \sum_{i=0}^{N} P_k(x_i)^2 = 0 \Leftrightarrow P_k(x_i) = 0 \; \forall i = 0, 1, \dots, N.$$

But this is impossible because P_k is a polynomial of degree $k < N + 1$, which would imply that P_k is zero. We will go through the proof step by step.

For $k = 1$, the polynomial $P_1(x) = \begin{vmatrix} s_0 & s_1 \\ 1 & x \end{vmatrix} = s_0 x - s_1$ is not identically zero. Therefore, $S_1 > 0$, $D_1 = s_0 > 0$ and $S_1 = D_1 D_2 > 0$, which implies that $D_2 > 0$.

$P_2(x)$ has the term $D_2 x^2$, which is not identically zero. Therefore, $S_2 > 0$. But, $S_2 = D_2 D_3$, so $D_3 > 0$. Repeating this argument shows that $D_{N+1} > 0$. We have to stop here, as it is possible to have $P_{N+1}(x_i) = 0$ for all $i = 0, 1, \dots, N$ without $P_{N+1}(x)$ being identically zero.

Returning to the minimization problem, equation [10.10] may be written as:

$$\underbrace{\begin{pmatrix} s_0 & s_1 & \dots & s_m \\ \vdots & \vdots & \vdots & \vdots \\ s_m & s_{m+1} & \dots & s_{2m} \end{pmatrix}}_{M_m} \begin{pmatrix} a_0 \\ \vdots \\ a_m \end{pmatrix} = \begin{pmatrix} t_0 \\ \vdots \\ t_m \end{pmatrix}, \qquad [10.11]$$

and $\det M_m = D_{m+1} > 0$ for all $m < N$. Therefore, the problem in [10.10] has a unique solution.

In addition to the condition that the first derivatives must vanish ($\frac{\partial S}{\partial a_k} = 0 \; \forall k = 0, \dots, m$), a sufficient condition for a point $a_0, a_1, \dots, a_m$ to minimize a function is for each of the determinants:

$$A_n = \begin{vmatrix} S_{a_0 a_0} & S_{a_0 a_1} & \dots & S_{a_0 a_n} \\ S_{a_1 a_0} & S_{a_1 a_1} & \dots & S_{a_1 a_n} \\ \vdots & \vdots & \vdots & \vdots \\ S_{a_n a_0} & S_{a_n a_1} & \dots & S_{a_n a_n} \end{vmatrix}$$

to be strictly positive for $n = 0, 1, \dots, m$, where $S_{a_j a_k}$ denotes the partial derivative $\frac{\partial^2 S}{\partial a_j \partial a_k}$.

It can be verified that:

$$\frac{\partial^2 S}{\partial a_j \partial a_k} = 2 \sum_{i=0}^{N} x_i^{j+k} = 2 S_{j+k},$$

and, therefore, $A_n = 2^{n+1} D_{n+1} > 0$ for $n \leq N$.

There is one single minimum (global minimum).

10.3. Algebraic formulation

We will consider the same problem as in the previous section, but this time from an algebraic perspective. Suppose that $Y \in \mathbb{R}^m$ and $A \in \mathcal{M}_{mn}(\mathbb{R})$ are given. We need to find an element $\hat{X} \in \mathbb{R}^n$ that makes $\|AX - Y\|_2$ as small as possible. We will see how this least-squares problem can be interpreted in terms of orthogonal projections and how it can be solved algebraically.

10.3.1. *Standard results on orthogonality*

This section reviews some results on orthogonality in Euclidean spaces. Below, we will write $< .|. >$ for the Euclidean scalar product in $\mathbb{R}^m$ ($< X|Y >= Y^T X$).

THEOREM.– *Every finite-dimensional Euclidean space admits an orthonormal basis.*

THEOREM.– *Let E be an n-dimensional vector subspace of $\mathbb{R}^m$. Then,*

– there exists a vector subspace $E^\perp$, known as the orthogonal complement of E, such that:

$$\forall X \in E, \forall Y \in E^\perp, \quad < X|Y >= 0;$$

– $\mathbb{R}^n = E \oplus E^\perp$.

This result follows directly from Schmidt orthogonalization.

LEMMA.– *Let $Y \in \mathbb{R}^m$. There exists a unique $\hat{Y} \in E$ such that:*

$$Y - \hat{Y} \in E^\perp.$$

The vector $\hat{Y}$ is called the orthogonal projection of Y onto the vector subspace E.

THEOREM.– *Suppose that the vectors $X, Y \in \mathbb{R}^m$ satisfy $< X|Y >= 0$. Then:*

$$\|X + Y\|_2 = \|X\|_2^2 + \|Y\|_2^2.$$

THEOREM.– *Let $A \in \mathcal{M}_{nm}(\mathbb{R})$ and $X \in \mathbb{R}^m$. Then:*

$$X \in Im(A)^\perp \leftrightarrow X \in Ker(A^T).$$

10.3.2. *Least-squares problem*

Suppose that $Y \in \mathbb{R}^m$ is given. Returning to the problem that we considered earlier, we need to find $\hat{X} \in \mathbb{R}^n$ such that:

$$\|A\hat{X} - Y\|^2 \leq \|AX - Y\|^2 \quad \forall X \in \mathbb{R}^n. \qquad [10.12]$$

Suppose that $\text{rank}(A) = n$, i.e. that the solution is unique. Let $E = \text{Im}(A)$. Problem [10.12] can be divided into two steps:

1) find $\hat{Y} \in E$ such that $\|\hat{Y} - Y\|^2 \leq \|Z - Y\|^2 \quad \forall Z \in E$;

2) find $\hat{X}$ such that $\hat{Y} = A\hat{X}$.

To solve the first step, we can decompose Y into $Y = \hat{Y} + \tilde{Y}$, where $\hat{Y} \in E$ and $\tilde{Y} \in E^{\perp}$. Indeed, given any arbitrary $Z \in E$, we can write $Z = \hat{Y} + \hat{Z}$ with $\hat{Z} \in E$. If $\hat{Z} \neq 0$, it follows that:

$$\begin{aligned} \|Z - Y\|^2 &= \|\hat{Y} - Y + \hat{Z}\|^2 \\ &= \|\hat{Y} - Y\|^2 + \|\hat{Z}\|^2 \; > \; \|\hat{Y} - Y\|^2. \end{aligned}$$

REMARK.–

$$\left.\begin{array}{r} Y - \hat{Y} \in (\mathrm{Im}(A))^{\perp} = \mathrm{Ker}(A^T) \Leftrightarrow A^T(Y - \hat{Y}) = 0 \\ \hat{Y} = A\hat{X} \end{array}\right\} \Leftrightarrow A^TY = A^TA\hat{X}.$$

10.3.3. *Solving by orthogonalization*

The two steps described above can be implemented as follows:

1) Using the Schmidt orthogonalization algorithm, find an orthogonal basis $\{q^1, \dots, q^n\}$ of $\mathrm{Im}(A)$.

2) Find the components of Y with respect to this basis and calculate its orthogonal projection $\hat{Y}$ onto $\mathrm{Im}(A)$.

By writing Q for the matrix whose columns are the vectors q^i, it can be shown that:

$$\hat{Y} = QQ^TY.$$

3) Calculate $\hat{X}$ by solving the linear system:

$$A\hat{X} = \hat{Y}. \qquad [10.13]$$

REMARK.– To solve the system in [10.13], we need to reduce it to a system of n equations and n unknowns. This can be made easier by taking advantage of the orthogonality properties of the basis $\{q^1, \dots, q^n\}$, and therefore of the columns of the matrix Q ($Q^TQ = I$) as follows:

$$Q^TA\hat{X} = Q^T\hat{Y} = Q^TY. \qquad [10.14]$$

Note also that Q^TA is an upper triangular square matrix.

NUMERICAL CONSIDERATIONS.– We now have two possible ways to solve a least-squares problem, but neither approach is particularly efficient:

– it is somewhat heavy-handed to attempt to solve the normal equations directly. In general, the system will be ill-conditioned;

– the linear system [10.14], constructed by the orthogonalization method, tends to be better conditioned. However, Schmidt orthogonalization is numerically imprecise and liable to accumulate rounding errors, so we cannot use it. The underlying idea of orthogonal projections can still be used as described below.

10.4. Numerically solving linear equations by QR factorization

10.4.1. *Householder transformations*

Note that the image of an orthonormal family under an orthogonal transformation is an orthonormal family. In the Euclidean vector space $\mathbb{R}^n$, orthogonal transformations preserve the value the Euclidean norm.

DEFINITION.– *Let $v \in \mathbb{R}^n$, $v \neq 0$. A Householder matrix is a matrix H of the form:*

$$H = I - 2\frac{vv^t}{v^t v}. \qquad [10.15]$$

THEOREM.– *Every Householder matrix is symmetric and orthogonal.*

We can note that:

$$Hv = I - \frac{2}{v^T v} vv^T \; v = -v,$$

and, for any $X \in \mathbb{R}^n$, if $< X|v >= 0$, then $HX = X$. Therefore, the matrix H represents a reflection about the hyperplane orthogonal to v.

THEOREM.– *Let X and Y be any two non-collinear normalized vectors in $\mathbb{R}^n$. There exists a Householder matrix H such that $HX = Y$.*

10.4.2. *QR factorization*

Let $A \in \mathcal{M}_{mn}(\mathbb{R})$ for $m \geq n$. There exists an orthogonal matrix $Q \in \mathcal{M}_{mm}(\mathbb{R})$ and an upper triangular matrix $R \in \mathcal{M}_{nn}(\mathbb{R})$ such that:

$$A = Q\begin{pmatrix} R \\ 0 \end{pmatrix}. \qquad [10.16]$$

REMARK.– Much like Schmidt orthogonalization, QR factorization orthogonalizes the columns of the matrix A. Numerically, it is much less prone to rounding errors.

10.4.3. *Application to the least-squares problem*

We need to minimize the function $E(X) = \|AX - Y\|_2^2$. Suppose that we have factorized the matrix A into:

$$A = Q\begin{pmatrix} R \\ 0 \end{pmatrix},$$

where $R \in \mathcal{M}_{nn}(\mathbb{R})$ is upper triangular and $Q \in \mathbb{M}_{mm}(\mathbb{R})$ is orthogonal. Then:

$$\|AX - Y\|^2 = \|Q^t AX - Q^T Y\|^2$$

(because the Euclidean norm is preserved by orthogonal transformations). Hence, if we decompose the matrix Q (by blocks) into:

$$Q = (Q_1 \, Q_2),$$

where $Q_1 \in \mathcal{M}_{mn}(\mathbb{R})$ and $Q_2 \in \mathcal{M}_{m,m-n}(\mathbb{R})$, then:

$$Q^T AX - Q^T Y = \begin{pmatrix} RX - Q_1^T Y \\ Q_2^T Y \end{pmatrix},$$

which implies that:

$$\|AX - Y\|^2 = \|RX - Q_1^T Y\|^2 + \|Q_2^T Y\|^2.$$

The second term in this sum is constant, so the vector $\hat{X}$ that minimizes the function $E(X)$ is a solution of the linear system:

$$R\hat{X} = Q_1^T Y. \quad [10.17]$$

This solution is unique if R is invertible, i.e. if rank$(A) = n$.

REMARK.– The linear systems [10.14] and [10.17] are identical. Only the methods used to compute the matrix and the right-hand side of the equation differ.

10.5. Applications

The method of least-squares has two major applications to problems with discrete data.

10.5.1. *Curve fitting*

By using the polynomial:

$$p(x) = a_0 + a_1 x + \cdots + a_m x^m$$

to model the data points $p(x)$, we can replace the original (possibly irregular) data-generating function by a straight line, parabola or any other type of fitted curve. The degree of $p(x)$ depends on the circumstances. Least-squares parabolas on five points (x_i, y_i) for $i = k-2, k-1, \ldots, k+2$ are a common choice, leading to the fitting equation:

$$y(x_k) \approx p(x_k) = y_k - \frac{3}{35}\delta^4 y_k.$$

This formula mixes together the five values $y_{k-2}, \ldots, y_{k+2}$ to generate a new estimate of the exact but unknown value $y(x_k)$. This formula must be modified slightly at the end points of finite data sets.

The square root of the average quadratic distance from the approximation set A_i to the corresponding exact values T_i (also known as the standard deviation or the square root of the variance) may be written as:

$$\text{standard deviation} = \left[\sum_{i=0}^{N} \frac{(T_i - A_i)^2}{N}\right]^{1/2}.$$

In test cases where T_i are known, we can use this parameter as a metric to estimate the quality of the fitting generated by the method of least-squares.

10.5.2. *Approximations of derivatives*

In practice, fitting an interpolating polynomial to a set of irregular data produces very poor estimates of derivatives. Even insignificant errors can be magnified until they dominate. Least-squares polynomials, on the contrary, do not interpolate; instead, they insert themselves in between the data points. This results in a smoothed function that typically provides better estimates of the derivatives, i.e. the values of $p'(x)$. The five-point parabola mentioned above leads to the expression:

$$y'(x_k) \approx p'(x_k) = \frac{1}{10h}(-2y_{k-2} - y_{k-1} + y_{k+1} + 2y_{k+2}).$$

This formula must also be modified slightly at the end points of the data set. In general, it leads to considerably better results than those obtained by differentiating interpolating polynomials. However, using these values with $p'(x_k)$ to attempt to estimate $y''(x_k)$ will once again lead to very unreliable results.

10.6. Using *Matlab*

The command p =*polyfit*(x,y,n) returns the polynomial p of degree n that approximates the curve $y = f(x)$ in the sense of least-squares.

To find the error between the experimental curve and the obtained model, we can use the function *polyval*(p,x), which returns the value of the polynomial p at each element of the vector or matrix x.

The command s= *std*(x) returns the standard deviation of a vector x. If x is a matrix, s is computed for each column of x. The command xbar = *mean*(x) computes the average value of the elements of x. If x is a matrix, xbar is computed for each column of x.

The following script computes the coefficients of the least-squares approximating polynomial using the normal equations. To solve the normal equations, it uses the *gaussPivot* function presented in Chapter 7. This script returns a polynomial of degree $m - 1$ that smooths a set of n experimental data points.

```
function coeff = polynFit(xData,yData,m)
% calculates the coefficients of the polynomial
% a(1)*x^(m-1) + a(2)*x^(m-2) + ... + a(m)
% that smooths the data points in the sense of least squares
% USE: coeff = polynFit(xData,yData,m)
% xData = x-coordinates of the data
% yData = y-coordinates of the data
A = zeros(m); b = zeros(m,1); s = zeros(2*m-1,1);

for i = 1:length(xData)
temp = yData(i);
for j = 1:m

b(j) = b(j) + temp;
temp = temp*xData(i);
end
temp = 1;

for j = 1:2*m-1
s(j) = s(j) + temp;
temp = temp*xData(i);
end
end
for i = 1:m
for j = 1:m
A(i,j) = s(i+j-1);
end
end
% Rearrange the coefficients so that the coefficient
% of x(m-1) is first
coeff = flipdim(gaussPiv(A,b),1)
```

After computing the coefficients of the polynomial, the standard deviation σ can be calculated with the following script:

```
function sigma = stdDev(coeff,xData,yData)
% returns the standard deviation between the data
% and the polynomial
% a(1)*x^(m-1) + a(2)*x^(m-2) + ... + a(m)
% coeff = coefficients of the polynomial
% xData = x-coordinates of the data
% yData = y-coordinates of the data
m = length(coeff);
n = length(xData); sigma = 0;
```

```
for i =1:n

y = polyEval(coeff,xData(i));
 sigma = sigma + (yData(i) - y)^2;
end
sigma =sqrt(sigma/(n - m));

function y = polyEval(coeff,x)
% Returns the values of the polynomial at x.
m = length(coeff);
y = coeff(1);
for j = 1:m-1
y = y*x + coeff(j+1);
end
```

The next script computes the variance of a vector using the two functions *polynFit* and *stdDev*:

```
xData=[-0.04,0.93,1.95,2.90,3.83,5.0,5.98,7.05,8.21,9.08,10.09]';
yData=[-8.66,-6.44,-4.36,-3.27,-0.88,0.87,3.31,4.63,6.19,7.4,8.85]';
format short e
while 1
k = input('degree of the polynomial  = ');

if isempty(k)
fprintf('done')
break
end

coeff = polynFit(xData,yData,k+1)
sigma = stdDev(coeff,xData,yData)
fprintf('\n')
end
```

Consider the following experimental data:

x	-0.04	0.93	1.95	2.90	3.83	5.00	5.98	7.05	8.21	9.08	10.09
y	-8.66	-6.44	-4.36	-3.27	-0.88	0.87	3.31	4.63	6.19	7.40	8.85

We can use the scripts listed above to compute the coefficients of the polynomial and the variance. After running these scripts, we find the following results for $k = 1$ and $k = 2$:

```
degree of the polynomial  = 1

coeff =
  1.7286e+000
 -7.9453e+000

sigma =
  5.1128e-001

degree of the polynomial  = 2

coeff =
 -4.1971e-002
  2.1512e+000
 -8.5701e+000

sigma =
  3.1099e-001
```

PART 4

Appendices

Appendix 1

Introduction to *Matlab*

A1.1. Introduction

Matlab stands for matrix laboratory. Originally written in Fortran by Cleve Moler, *Matlab* was designed to make it easier to access the matrix software developed by the LINPACK and EISPACK projects. The modern version of *Matlab* is written in C, published by MathWorks Inc. and is available in both professional and student versions across multiple platforms.

Matlab is a powerful, comprehensive and easy-to-use environment for scientific computation. It gives engineers, researchers and scientists an interactive system that integrates numerical calculations and visualizations.

Matlab uses its own intuitive and natural programming language, which offers spectacular CPU performance improvements over other languages such as C, TurboPascal and Fortran. *Matlab* allows users to dynamically include links to programs in C or Fortran, exchange data with other software applications or use *Matlab* itself as the engine for analysis and visualization.

Matlab also offers specialized tools for certain fields, known as ToolBoxes, which are regarded as one of *Matlab*'s most attractive features for most users. These ToolBoxes are collections of functions that extend the *Matlab* environment to enable it to solve certain types of problems. They cover a wide range of topics, including signal processing, automation, neural networks, structural computations and statistics.

Matlab allows users to work interactively in either command mode or programming mode; graphical visualizations can be generated in either case. Widely viewed as one of the best programming languages (alongside others such as C or Fortran), *Matlab* offers the following specific advantages relative to its competitors:

– easy programming;

– continuity between integer, real and complex values;

– an extensive range of numbers and precisions;

– a very comprehensive mathematical library;

– graphical tools, including graphical interface tools and utilities;

– the ability to add links to other classical programming languages (e.g. C or Fortran).

The graphical interface allows scientific or even artistic representations to be generated from their mathematical expressions. The figures generated by *Matlab* are simple and eye-catching, and an impressive array of features is available to enhance them.

A1.2. Starting up *Matlab*

To launch *Matlab*:

– in Windows, click on Start, then Programs, then *Matlab*;

– for other operating systems, refer to the user manual.

The *Matlab* prompt ">>" should then be displayed. This is where users can enter commands (see Figure A1.1).

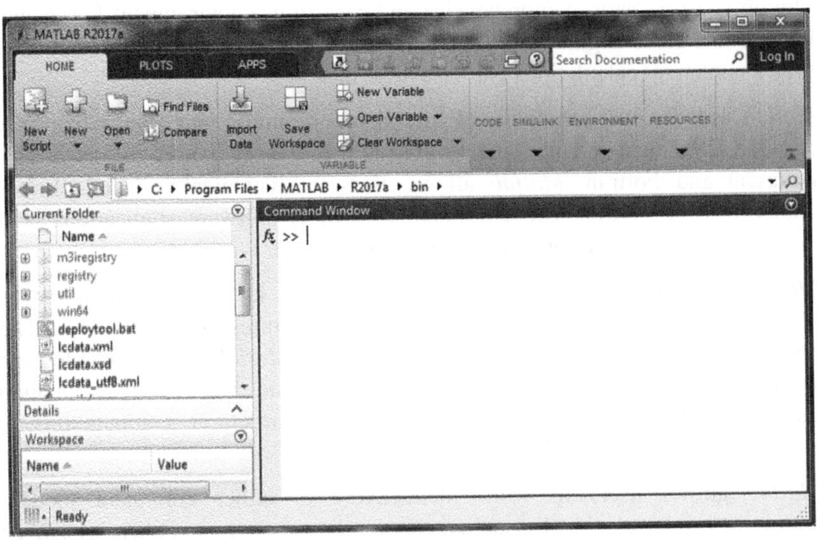

Figure A1.1. *Matlab window*

The *quit* function allows the user to exit *Matlab*:

```
>>quit
```

The "help" command gives information about specific problems. For example:

```
>> help cos

COS Cosine.
COS(X) is the cosine of the elements of X.
```

The standard arithmetic operations are supported:

+ : addition; - : subtraction; / : division; * : multiplication; ^ : exponentiation; pi=π. For example:

```
>>x=2

x =
      2
>>P=(4*x^2-2*x+3)/(x^3+1)

P =
      1.6667
```

Command mode allows us to perform computations in *Matlab*.

For example, suppose that we wish to calculate the following volume: $V = \frac{4}{3}\pi R^3$, where $R = 4$ cm. We can carry this out as follows:

```
>>R=4

R =
      4
>>V=4/3*pi*R^3

V =
      268.0826
```

A1.3. Mathematical functions

The trigonometric functions are:

sin(x), cos(x), tan(x), asin(x), acos(x), atan(x), atan2(x,y), sinh(x), cosh(x), tanh(x), asinh(x), acosh(x), atanh(x).

The other (elementary) mathematical functions available in *Matlab* are:

abs(x)	Absolute value of x
angle(x)	If x is a positive real number, then angle(x) = 0; if x has positive imaginary part, then angle(x) = pi/2
sqrt(x)	Square root of x
real(x)	Real part of the complex value x
imag(x)	Imaginary part of the complex value x
conj(x)	Complex conjugate of x
round(x)	Round x to the nearest integer
fix(x)	Round x toward zero
floor(x)	Round x to the integer below
ceil(x)	Round x to the integer above
sign(x)	=+1 if x>0; =-1 if x<0
rem(x,y)	The remainder of the division: = x-y*fix(x/y)
exp(x)	Exponential (base e)
log(x)	Logarithm (base e)
log10(x)	Logarithm (base 10)

A1.4. Operators and programming with *Matlab*

Most programming scripts make frequent use of "if" statements. Every "if" statement must be followed by the "end" keyword:

```
>>V=268.0826

V =
    268.0826
>> R = 4

R =
    4
>>if V>150, surface=pi*R^2, end

surface =
    50.2655
```

The "not" operator is written (or symbolically represented) by "~=":

```
>>R=4
```

```
R =
    4
>>if R ~=2, V=4/3*pi*R^3, end

V=
    268.0826
```

The "equals" operator (==) in an "if" statement is written (or symbolically represented) by "==":

```
>>R=4
R =
    4
>>if R==4, V=4/3*pi*R^3; end
```

The "or" operator is written (or symbolically represented) by "|". For example, the test "if $R = 4$ or $m = 1$" can be written as:

```
>>if R==4 | m==1, V=4/3*pi*R^3; end
```

Other operators available in *Matlab* include:

>	greater than	<	less than
>=	greater than or equal to	<=	less than or equal to
^	power (exponent)	*	product
xor	exclusive OR (XOR)	/	division

Error displays the message: "error"

For example, "if $g > 2$ or $g < 0$, then $a = 4$" can be programmed as follows:

```
>>if g>2 |g<0,
a=4;
end
```

As another example, "if a > 3 and c < 0, then b = 15" can be implemented with the code:

```
>>if a>3 \& c<0,
b=15;
end
```

The operators "&" and "|" can be chained together:

```
>>if ((a==2 | b==3)\&(c<5)),
g=1;
end
```

The operator “if ... elseif ... else ... end” can be used as follows:

```
>>R=2, if R>3, b=1 ;
elseif R==3, b=2;
else b=0;
end
```

The keyword “elseif” can be repeated as many times as necessary within the same program.

The “for/end” and “while/end” operators can, for example, be used as follows:

```
>>for R=1 :5,
V=4/3*pi*R^3;
disp([R,V]);
end
```

In this example, R ranges from 1 to 5 and the command “disp([R,V])” returns the matrix [R=1 :5,V (V(1) :V(5)].

The “length” command can also be called, returning the size of a variable. In the above example, length(R)=5; (R=1 :5) and length(R)-1=4 (4 intervals separated by increments of 1).

```
>>while R<5, R=R+1 ; V=4/3*pi*R^3; disp([R,V]); end
```

The “while” command continues to execute the same instruction until its logical test no longer returns true.

The next example demonstrates how to use an automatic increment in a “for” loop:

```
>>for R=5 :-1 :1, V=4/3*pi*R^3; disp([R,V]); end
```

Here, the increment is decreasing (=-1). “For” loops can be nested as many times as necessary:

```
>>for i=0 :10, for j=1 :5, a=i*j; end; end
```

A1.5. Writing a *Matlab* script

Matlab programs are saved with the file extension “.m”. When they are executed, error messages are displayed whenever *Matlab* encounters an error, indicating where in the code the error arose. Scripts do not need to be precompiled. To run a program, we must first load the file directory in which it is saved.

Data files are saved with the file extension “.mat”, and variables are saved with double precision.

A1.6. Generating figures with *Matlab*

Matlab is a very powerful and user-friendly tool for managing figures, whether in one, two or three dimensions. For example, to plot the curve y=sin(x), where x=0:50, we can simply run:

```
>>x= -pi:pi/10:pi; y=sin(x), plot(x,y)
```

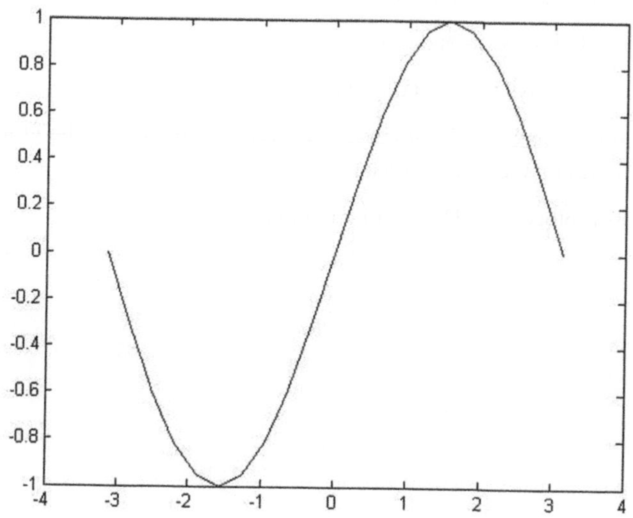

Some of the many commands for plotting graphs and manipulating axes and scales are briefly summarized below:

– xlabel("time"): gives a title to the x-axis;

– ylabel("speed"): gives a title to the y-axis;

– title("progression of the speed"): gives a title to the graph;

– text(2,4,"+++Temperature T1"): adds a caption to the curve plotted by "+++" at a certain point;

– loglog(x,y): plots the curve with a logarithmic scale (log-log);

– semilogx(t,f(t)): plots the curve with a logarithmic scale along the x-axis only;

– semilogy(t,f(t)): plots the curve with a logarithmic scale along the y-axis only;

– grid on: displays the grid in the graph;

– grid off: hides the grid in the graph;

– clf: deletes the graph;

– plot(x,y,x,z,x,w): plots y, z and w as a function of x on the same graph;

– polar(x,y): plots the curve y as a function of x in polar coordinates;

– plot(x,y,“+g”): plots y as a function of x using green “+” symbols;

– fplot(“f_name”,[x-min, x-max]): plots the function “f_name” on the specified interval of x;

– axis(“square”): plots a square graph;

– axis(“off”): hides the x- and y-axes;

– axis(“on”): displays the x- and y-axes;

– axis([x-min, x-max, y-min, y-max]): displays the section of the graph between the specified values on the x- and y-axes;

– plot3(x,y,z): plots z in 3D as a function of x and y.

Appendix 2

Introduction to Optimization

A2.1. Introduction

The term "optimization" usually refers to the act of minimizing or maximizing a function. This can be reduced to the problem of minimizing an arbitrary function $f(x)$, because maximizing $f(x)$ is equivalent to minimizing $-f(x)$. In engineering, optimization is often performed during the design phase. The function $f(x)$ is said to be the cost function or the objective function and represents some quantity that is undesirable, such as the cost or the weight. The components of the vector x are called the design variables and can be varied. Physical dimensions (e.g. lengths, areas, angles) are common examples of design variables.

This chapter briefly introduces the field of optimization by giving the precise statement of the problem of optimization without constraints. An extensive range of literature is available on the subject of optimization. We will also present the techniques most commonly used to find the minimum of a function.

Minimization algorithms are iterative procedures that require an initialization vector x^0. If the function $f(x)$ has multiple local minima, the choice of x^0 determines the local minimum that will be found by the algorithm. We cannot guarantee that we will find the global minimum. One approach is to compute several local minima and select the one which gives the best value for the function $f(x)$.

A2.2. Standard results on functions from $\mathbb{R}^n$ to $\mathbb{R}$

NOTATION.–

– $x \in \mathbb{R}^n$ is a *column* vector with components x_i $i = 1, \ldots, n$.

– We will be working in the space $\mathbb{R}^n$, equipped with the Euclidean scalar product:

$$< x|y >= \sum_{i=1}^{n} x_i y_i$$

and the Euclidean norm:

$$\|x\| = \sqrt{< x|x >} = \left(\sum_{i=1}^{n} x_i^2\right)^{\frac{1}{2}}.$$

A2.2.1. *Differentiability*

DEFINITION.– *Let* $f\colon \mathbb{R}^n \to \mathbb{R}$. *We say that* f *is continuous at the point* x_0 *if:*

$$\forall \epsilon > 0,\ \exists \eta > 0 / \ \forall x \in \mathbb{R}^n,\ \|x - x_0\| \leq \eta \Rightarrow |f(x) - f(x_0)| \leq \epsilon.$$

If f *is continuous at* x_0 $\forall x_0 \in \mathbb{R}^n$, *then we say that* f *is continuous on* $\mathbb{R}^n$.

DEFINITION.– *Let* $f\colon \mathbb{R}^n \to \mathbb{R}$. *We say that* f *is differentiable at the point* $x \in \mathbb{R}^n$ *if there exists a vector* $g(x) \in \mathbb{R}^n$ *such that, for* $h \in \mathbb{R}^n$*:*

$$f(x+h) = f(x) + < g(x)|h > + \|h\|\epsilon(\|h\|),$$

where $\epsilon\ :\ \mathbb{R} \to \mathbb{R}$ *is a continuous function satisfying* $\lim\limits_{\lambda \to 0} \epsilon(\lambda) = 0$.

NOTATION.– The gradient vector of f at x is defined as a column vector, denoted $g(x)$ or $\vec{grad} f(x)$.

REMARK.– In the mathematical literature, the gradient of f at the point x is usually defined as the row vector $\nabla f(x)$ such that:

$$\nabla f(x) = g^T(x).$$

With this convention, the differentiability condition can be written as:

$$f(x+h) = f(x) + \nabla f(x)h + \|h\|\epsilon(\|h\|),$$

where $\nabla f(x)h$ is a matrix product.

DEFINITION.– *Consider the function* $f\colon \mathbb{R}^n \to \mathbb{R}$ *(not necessarily assumed to be continuous). Let* $x,\ y \in \mathbb{R}^n$. *We say that* f *admits a directional derivative at the point* x *in the direction* y *if the following limit exists:*

$$\lim_{\lambda \to 0, \lambda \neq 0} \frac{f(x + \lambda y) - f(x)}{\lambda}.$$

NOTATION.– This limit, if it exists, is called the directional derivative of f at x in the direction y, and is written $Df(x; y)$.

REMARK.– If f is differentiable at x, then it has directional derivatives in every direction at x. Partial derivatives are special cases of directional derivatives.

THEOREM.– *Let f be a function that is differentiable at $x \in \mathbb{R}^n$. Let $y \in \mathbb{R}^n$ be an arbitrary vector. Then:*

$$Df(x; y) = \nabla f(x)y.$$

THEOREM.– *The gradient vector is the vector of partial derivatives:*

$$\forall i \in \{1, \dots, n\}, \quad g_i(x) = \frac{\partial f}{\partial x_i}(x).$$

DEFINITION.– *The Hessian matrix (or just the Hessian) of f at x is defined as the matrix of second partial derivatives of f at x, and is written $\nabla^2 f(x)$. Thus:*

$$\left[\nabla^2 f(x)\right]_{ij} = \frac{\partial^2 f}{\partial x_i \partial x_j}(x).$$

Below, unless otherwise stated, we will assume that the functions we are working with are twice continuously differentiable, which guarantees that the Hessian matrix is well defined and symmetric.

A2.2.2. *Convexity*

DEFINITION.– *We say that $f\colon \mathbb{R}^n \to \mathbb{R}$ is convex if:*

$$\forall x, y \in \mathbb{R}^n,\ \forall \theta \in [0, 1] \quad f(\theta x + (1-\theta)y) \leq \theta f(x) + (1-\theta)f(y).$$

We say that f is strictly convex if:

$$\forall x, y \in \mathbb{R}^n,\ x \neq y,\ \forall \theta \in]0, 1[\quad f(\theta x + (1-\theta)y) \leq \theta f(x) + (1-\theta)f(y).$$

REMARK.– Case of functions in a single real variable. Let $f\colon \mathbb{R} \to \mathbb{R}$. The following equivalences describe the logical connections between convexity and positivity of the second derivative:

$$f \text{ convex} \Leftrightarrow f'' \geq 0,\ \forall x$$

$$f \text{ strictly convex} \Leftrightarrow f'' > 0,\ \forall x.$$

These properties can be generalized to arbitrary dimensions as statements about the properties of the Hessian.

A2.3. Optimization without constraints

A2.3.1. *Quadratic functions*

DEFINITION.– *A quadratic function is a function of the form:*

$$f(x) = \frac{1}{2}x^T Ax - b^T x, \qquad \text{[A2.1]}$$

where $A \in \mathcal{M}_{nn}(\mathbb{R})$ is a symmetric matrix and $b \in \mathbb{R}^n$.

THEOREM.– *The quadratic function defined by [A2.1] is strictly convex if and only if A is positive-definite.*

REMARK.– The function $f\colon \mathbb{R}^n \to \mathbb{R}$ is convex (respectively strictly convex) if and only if the Hessian $\nabla^2 f(x)$ is positive semi-definite (respectively positive-definite) for every $x \in \mathbb{R}^n$.

A2.3.2. *Existence and uniqueness results*

This section presents a framework in which optimization problems always have at least one solution [MIN 83].

THEOREM.– *Let $f\colon \mathbb{R}^n \to \mathbb{R}$ be a continuous and coercive*[1] *function. Then, there exists at least one element $\bar{x} \in \mathbb{R}^n$ such that:*

$$f(\bar{x}) \le f(x), \quad \forall x \in \mathbb{R}^n.$$

THEOREM.– *Let $K \subset \mathbb{R}^n$ be a closed set and suppose that $f\colon K \to \mathbb{R}$ is a continuous function. Suppose further that:*

$$\begin{cases} K \textit{ bounded} \\ or \\ f(x) \to +\infty \textit{ whenever } \|x\| \to \infty, \; x \in K \end{cases}$$

Then, there exists at least one element $\vec{x} \in \mathbb{R}$ such that:

$$f(\bar{x}) \le f(x), \quad \forall x \in K.$$

THEOREM.– *Let K be a convex subset of $\mathbb{R}^n$. Suppose that $f\colon K \to \mathbb{R}$ is a strictly convex function. Then, the minimum $\bar{x} \in K$, if it exists, is unique.*

REMARK.– The properties of convexity or strict convexity rarely hold in practice. We therefore need to restrict attention to the "local" behavior of the function. We say that $\bar{x}$ is a local minimum (see Figure A2.1) if there exists $\rho > 0$ such that:

$$\|x - \bar{x}\| \le \rho \Rightarrow f(\bar{x}) \le f(x).$$

A2.3.3. *Optimality conditions*

The discussion below only applies to optimization problems without constraints. Suppose that the function $f : \mathbb{R}^n \to \mathbb{R}$ is known and is continuously differentiable. The statement of the optimization problem is to find $\tilde{x} \in \mathbb{R}^n$ such that:

$$f(\tilde{x}) \le f(x), \quad \forall x \in \mathbb{R}^n. \qquad [A2.2]$$

1 We say that f is coercive if $f(x) \to +\infty$ whenever $\|x\| \to \infty$.

THEOREM.– *A necessary condition for the vector $\tilde{x}$ to be a solution of the optimization problem [A2.2] is given by*

$$\nabla f(\tilde{x}) = 0.$$

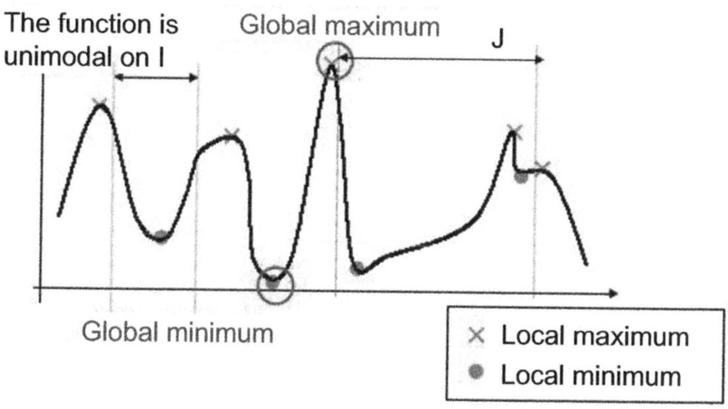

Figure A2.1. *Illustration of global and local minima*

REMARK.–

– This theorem gives a necessary condition that is not sufficient in general.

– This necessary condition also applies to local minima.

THEOREM.– *Suppose that f is a convex function. Let $\bar{x} \in \mathbb{R}^n$ such that $\nabla f(\bar{x}) = 0$. Then, $\bar{x}$ is a solution of the global optimization problem [A2.2].*

EXAMPLE.– Let $A \in \mathcal{M}_{nn}(\mathbb{R})$ be a symmetric matrix. Suppose that $b \in \mathbb{R}^n$. Consider the quadratic function f defined by $f(x) = \frac{1}{2}x^T Ax - b^T x$. The necessary condition for optimality may be stated as:

$$\nabla f(\bar{x}) = 0 \Leftrightarrow (A\bar{x} - b)^T = 0.$$

Therefore, if A is positive-definite, its unique minimum satisfies the equation:

$$A\bar{x} = b.$$

The problem of minimizing a quadratic function can therefore be reduced to the problem of solving a linear system.

Bibliography

[JED 01] JEDRZEJEWSKI F., *Introduction aux méthodes numériques*, Springer, 2001.

[KIU 05] KIUSALAAS J., *Numerical methods in engineering with MATLAB®*, Cambridge University Press, 2005.

[LAS 94a] LASCAUX P., THÉODOR R., *Analyse numérique matricielle appliquée á l'art de l'ingénieur*, vol. 2, Dunod, 1994.

[LAS 94b] LASCAUX P., THÉODOR R., *Analyse numérique matricielle appliquée á l'art de l'ingénieur*, vol. 1, Dunod, 1994.

[MIN 83] MINOUX M., *Programmation mathématique: Théorie et Algorithmes*, Eyrolles, Paris, 1983.

[ORT 70] ORTEGA J., RHAINBOLDT W., *Iterative Solution of Nonlinear Equations in Several Variables*, Academic Press, New York, 1970.

Index

A, B, C

D, E, F

G, H, I

J, K, L, M

N, O, P

R, S, T, V

Other titles from

iSTE

in

Mechanical Engineering and Solid Mechanics

2018

BOREL Michel, VÉNIZÉLOS Georges
Movement Equations 4: Equilibriums and Small Movements
(Non-deformable Solid Mechanics Set – Volume 4)

SALENÇON Jean
Virtual Work Approach to Mechanical Modeling

2017

BOREL Michel, VÉNIZÉLOS Georges
Movement Equations 2: Mathematical and Methodological Supplements
(Non-deformable Solid Mechanics Set – Volume 2)
Movement Equations 3: Dynamics and Fundamental Principle
(Non-deformable Solid Mechanics Set – Volume 3)

BOUVET Christophe
Mechanics of Aeronautical Solids, Materials and Structures
Mechanics of Aeronautical Composite Materials

BRANCHERIE Delphine, FEISSEL Pierre, BOUVIER Salima, IBRAHIMBEGOVIC Adnan
From Microstructure Investigations to Multiscale Modeling: Bridging the Gap

CHEBEL-MORELLO Brigitte, NICOD Jean-Marc, VARNIER Christophe
From Prognostics and Health Systems Management to Predictive Maintenance 2: Knowledge, Traceability and Decision
(Reliability of Multiphysical Systems Set – Volume 7)

EL HAMI Abdelkhalak, RADI Bouchaib
Dynamics of Large Structures and Inverse Problems
(Mathematical and Mechanical Engineering Set – Volume 5)
Fluid-Structure Interactions and Uncertainties: Ansys and Fluent Tools
(Reliability of Multiphysical Systems Set – Volume 6)

KHARMANDA Ghias, EL HAMI Abdelkhalak
Biomechanics: Optimization, Uncertainties and Reliability
(Reliability of Multiphysical Systems Set – Volume 5)

LEDOUX Michel, EL HAMI Abdelkhalak
Compressible Flow Propulsion and Digital Approaches in Fluid Mechanics
(Mathematical and Mechanical Engineering Set – Volume 4)
Fluid Mechanics: Analytical Methods
(Mathematical and Mechanical Engineering Set – Volume 3)

MORI Yvon
Mechanical Vibrations: Applications to Equipment

2016

BOREL Michel, VÉNIZÉLOS Georges
Movement Equations 1: Location, Kinematics and Kinetics
(Non-deformable Solid Mechanics Set – Volume 1)

BOYARD Nicolas
Heat Transfer in Polymer Composite Materials

CARDON Alain, ITMI Mhamed
New Autonomous Systems
(Reliability of Multiphysical Systems Set – Volume 1)

DAHOO Pierre Richard, POUGNET Philippe, EL HAMI Abdelkhalak
Nanometer-scale Defect Detection Using Polarized Light
(Reliability of Multiphysical Systems Set – Volume 2)

DE SAXCÉ Géry, VALLÉE Claude
Galilean Mechanics and Thermodynamics of Continua

DORMIEUX Luc, KONDO Djimédo
Micromechanics of Fracture and Damage
(Micromechanics Set – Volume 1)

EL HAMI Abdelkhalak, RADI Bouchaib
Stochastic Dynamics of Structures
(Mathematical and Mechanical Engineering Set – Volume 2)

GOURIVEAU Rafael, MEDJAHER Kamal, ZERHOUNI Noureddine
From Prognostics and Health Systems Management to Predictive Maintenance 1: Monitoring and Prognostics
(Reliability of Multiphysical Systems Set – Volume 4)

KHARMANDA Ghias, EL HAMI Abdelkhalak
Reliability in Biomechanics
(Reliability of Multiphysical Systems Set –Volume 3)

MOLIMARD Jérôme
Experimental Mechanics of Solids and Structures

RADI Bouchaib, EL HAMI Abdelkhalak
Material Forming Processes: Simulation, Drawing, Hydroforming and Additive Manufacturing
(Mathematical and Mechanical Engineering Set – Volume 1)

2015

KARLIČIĆ Danilo, MURMU Tony, ADHIKARI Sondipon, MCCARTHY Michael
Non-local Structural Mechanics

SAB Karam, LEBÉE Arthur
Homogenization of Heterogeneous Thin and Thick Plates

2014

ATANACKOVIC M. Teodor, PILIPOVIC Stevan, STANKOVIC Bogoljub, ZORICA Dusan
Fractional Calculus with Applications in Mechanics: Vibrations and Diffusion Processes

ATANACKOVIC M. Teodor, PILIPOVIC Stevan, STANKOVIC Bogoljub, ZORICA Dusan
Fractional Calculus with Applications in Mechanics: Wave Propagation, Impact and Variational Principles

CIBLAC Thierry, MOREL Jean-Claude
Sustainable Masonry: Stability and Behavior of Structures

ILANKO Sinniah, MONTERRUBIO Luis E., MOCHIDA Yusuke
The Rayleigh–Ritz Method for Structural Analysis

LALANNE Christian
Mechanical Vibration and Shock Analysis – 5-volume series – 3rd edition
Sinusoidal Vibration – Volume 1
Mechanical Shock – Volume 2
Random Vibration – Volume 3
Fatigue Damage – Volume 4
Specification Development – Volume 5

LEMAIRE Maurice
Uncertainty and Mechanics

2013

ADHIKARI Sondipon
Structural Dynamic Analysis with Generalized Damping Models: Analysis

ADHIKARI Sondipon
Structural Dynamic Analysis with Generalized Damping Models: Identification

BAILLY Patrice
Materials and Structures under Shock and Impact

BASTIEN Jérôme, BERNARDIN Frédéric, LAMARQUE Claude-Henri
Non-smooth Deterministic or Stochastic Discrete Dynamical Systems: Applications to Models with Friction or Impact

EL HAMI Abdelkhalak, RADI Bouchaib
Uncertainty and Optimization in Structural Mechanics

KIRILLOV Oleg N., PELINOVSKY Dmitry E.
Nonlinear Physical Systems: Spectral Analysis, Stability and Bifurcations

LUONGO Angelo, ZULLI Daniele
Mathematical Models of Beams and Cables

SALENÇON Jean
Yield Design

2012

DAVIM J. Paulo
Mechanical Engineering Education

DUPEUX Michel, BRACCINI Muriel
Mechanics of Solid Interfaces

ELISHAKOFF Isaac *et al.*
Carbon Nanotubes and Nanosensors: Vibration, Buckling and Ballistic Impact

GRÉDIAC Michel, HILD François
Full-Field Measurements and Identification in Solid Mechanics

GROUS Ammar
Fracture Mechanics – 3-volume series
Analysis of Reliability and Quality Control – Volume 1
Applied Reliability – Volume 2
Applied Quality Control – Volume 3

RECHO Naman
Fracture Mechanics and Crack Growth

2011

KRYSINSKI Tomasz, MALBURET François
Mechanical Instability

SOUSTELLE Michel
An Introduction to Chemical Kinetics

2010

BREITKOPF Piotr, FILOMENO COELHO Rajan
Multidisciplinary Design Optimization in Computational Mechanics

DAVIM J. Paulo
Biotribolgy

PAULTRE Patrick
Dynamics of Structures

SOUSTELLE Michel
Handbook of Heterogenous Kinetics

2009

BERLIOZ Alain, TROMPETTE Philippe
Solid Mechanics using the Finite Element Method

LEMAIRE Maurice
Structural Reliability

2007

GIRARD Alain, ROY Nicolas
Structural Dynamics in Industry

GUINEBRETIÈRE René
X-ray Diffraction by Polycrystalline Materials

KRYSINSKI Tomasz, MALBURET François
Mechanical Vibrations

KUNDU Tribikram
Advanced Ultrasonic Methods for Material and Structure Inspection

SIH George C. *et al.*
Particle and Continuum Aspects of Mesomechanics

Printed and bound by CPI Group (UK) Ltd, Croydon, CR0 4YY
16/09/2026
14963075-0003